Global Food Security and Supply

T0201337

Global Food Security and Supply

Wayne Martindale
Sheffield Business School
Sheffield Hallam University
UK

WILEY Blackwell

This edition first published 2015 © 2015 by John Wiley & Sons, Ltd

Registered office: John Wiley & Sons, Ltd, The Atrium, Southern Gate, Chichester, West Sussex, PO19 8SQ, UK

Editorial offices: 9600 Garsington Road, Oxford, OX4 2DQ, UK
The Atrium, Southern Gate, Chichester, West Sussex, PO19 8SQ, UK
111 River Street, Hoboken, NJ 07030-5774, USA

For details of our global editorial offices, for customer services and for information about how to apply for permission to reuse the copyright material in this book please see our website at website at www.wiley.com/wiley-blackwell.

Library of Congress Cataloging-in-Publication Data has been applied for.

ISBN: 978-1-1186-9932-4 (paperback)

A catalogue record for this book is available from the British Library.

Wiley also publishes its books in a variety of electronic formats. Some content that appears in print may not be available in electronic books.

Cover image: "World Map with Spices" © iStock/ susoy

Set in 12/14 PalatinoLTStd-Roman by Toppan Best-set Premedia Limited
Printed and bound in Malaysia by Vivar Printing Sdn Bhd

1   2015

For my wife, Deborah, and my daughters, Minnie, Tula, and Timéa, who will continue to make the world a better place.

Contents

About the Author

Dr Wayne Martindale is a Research Director for MPC Research and an Editor for 'Science Into', an on-line magazine for food industry innovations. He is a leading figure in the food and agricultural industry for sustainability issues. Dr. Martindale has been an Organisation for Economic Co–operation and Development (OECD) Cooperative Research Programme Fellow, British Grassland Society Fellow for sustainable agricultural systems, and a Commonwealth Scientific and Industrial Research Organisation (CSIRO) McMaster fellow for sustainable food processing. Starting his scientific career as a biochemist at the University of Sheffield he then applied research with the United States Department of Agriculture (USDA) and Levington Agriculture Ltd where he established the first independent fertiliser industry technical information service for chartered agronomists. He has developed undergraduate and postgraduate programmes across the agri-food supply chain at the universities of Leeds, Sheffield, and Sunderland in the United Kingdom. He is also a research fellow in Corporate Social Responsibility at Sheffield Business School.

See www.waynemartindale.com for his current research and updates.

Preface

For me, the route to publishing this book started after I had delivered a conference for the Organisation for Economic Co–operation and Development (OECD) Cooperative Research Programme at Royal Holloway and Bedford College at the University of London in September 2010. I was in the middle of an exciting McMaster Fellowship for the Commonwealth Scientific and Industrial Research Organisation (CSIRO) in Australia and the proceedings of the conference were published by the Association of Applied Biologists in the United Kingdom. All this activity led Andrew Harrison of Wiley to approach me to write a book on food security; it also led to me thinking I was so busy I would not go near such a project. After a few months of thinking about this and looking at the contributors to the proceedings of the London conference Mr. Harrison's insight and tenacity has paid off in that I have written the book in front of you.

The need for supply chain solutions to food security as always struck me as an essential part of getting to a secure and sustainable diet for 9 billion global citizens some 30 years into our future. Having worked in agriculture I was often surprised at a lack of connectivity to the food industry, and when I moved to work with the food industry I was shocked at the level of connectivity to agriculture. When the consumer is placed into this melting pot of conflicts there are naturally pressures and misunderstandings concerned with the sustainability of foods. I believe the food industry has responded to integrating supply chains; there were always examples of these practices but they have become common place even if they are often out of sight from the consumer. This book has tackled the issue of food security pressures operating at global levels and relates them to the operations of the food supply chain and how we put meals together to eat. The subject matter is wide-ranging and there is much I have left out of this book, but I hope to have achieved a balance of informing those who are interested, strengthening the understandings of those in the food industry, and allowing the reader to focus on sustainable solutions.

My book is aimed at those who have an understanding of how the food supply chain behaves and who want to know more. It will guide undergraduate students as well as the informed reader and established expert of food sustainability. I have strived to achieve a generalist approach while not losing understanding that is developed by my expert experience of teaching agriculture, sustainability, and food manufacturing to a wide range of students.

I would like to thank the team at Wiley who have efficiently guided the processing of my manuscript. I would also like to thank the people who have supported me and discussed several issues regarding the food security debate with me, including Adam Bedford at the National Farmers Union (NFU) in Brussels, Chandru Chandrasekhar of Sustein Ltd, Alan Marson of New Food Innovation Ltd, Professor Tim Benton at University of Leeds, Dr Murray Clark of Sheffield Business School, and Dr Martin L Warnes of Ipswich High School. This book is largely a product of the friendship and comradeship I have experienced and the sum of my curiosity that started growing up in Suffolk roaming the Stour Valley around Great Cornard and Sudbury. There are numerous key characters who helped my route to authoring this book, too numerous to mention, but Dr Ian Richards, who was Managing Director at Levington Agriculture Ltd; Professor Peter Lillford CBE, who was a Chief Scientist at Unilever; and Jay Sellahewa, at CSIRO, have had major parts to play in helping me to gain the confidence to put forward ideas and applications in this arena. Finally, a massive heartfelt thank goes to my wife, Deborah, and my children, Minnie, Tula, and Timéa Alred-Martindale, and also to Lloyd and Rose Ashton, who never stop inspiring me.

Dr Wayne Martindale
Barnsley, UK
September 2014

Introduction

A Reflection on Why We Should Care about Food Security

I am sitting on Oxley Bank in the United Kingdom, which is part of the Bretton Estate, which since 1977 has been the Yorkshire Sculpture Park and a deer or hunting park for a thousand years before that. It is a mid-August morning looking west out over a working landscape of West Yorkshire towards Huddersfield where, in the distance, the Emley Moor telecommunications mast, which is over 300 m in height, serves to mark where I am heading. In the foreground landscape, there is a patchwork of oilseed canola, forage maize, and pasture crops in various stages of production. A small herd of 30 dairy Holstein cows grazes one of the fields near me that is just beginning to 'burn-off', that is, to dry out and turn brown in the late summer sun. This landscape is an idyllic view of the countryside in the United Kingdom; there is relative quiet, even though the M1 motorway, one of our main arterial routes through the country, is 3 km behind me in a valley, with the Bretton Lake surrounded by beech trees. The landscape is working with dairying, tourism, intellectual wealth, arable farming, forestry, and wind turbines, and a cluster of small business workshops in the near distance provides a constant purr of metalworking machinery. This is a productive-landscape worked by agricultural activities that has moved with the times in European terms.

My thoughts are of being pleased to be here and this turns my attention to the food security debate because for one I have to return home and write this book, but also I have constantly questioned how real can the food security threat be since 1975, when I first saw an essay by Indira Gandhi, which I will refer to in this book.[1] From my vantage point here looking out of what is clearly a well-regulated and managed landscape that provides most of my needs for an acceptable quality of life, it seems an abstract issue. Here I sit on a rural earth bank in West Yorkshire topped by a wire frame sculpture that people climb through and think playful or intellectual thoughts in a landscape that provides work, food, and social capital, but the global food system is on a precipice of shortage and limitation. I am told this, and it is my job to make sense of data and statistics that can evidence stresses and strains in a world food system. What I have come to understand is that food is a very specialized consumer good in that it provides health and pleasure and its time as a food product is extremely transient. Relating this transience to high-level issues of land production, resource limitation, and global shortage of food is extremely difficult because ultimately we have to consider constraint in terms of reducing volumes consumed or reduction in access. This has been achieved in recent years by increases in food prices, which have focused discussions on food security in Europe as never before. The 1970s and 1980s hunger and security issues were driven by individuals such as Indira Gandhi, Willy Brandt, and Gro Harlem Brundtland, and evidence provided by commentators took a Malthus-like view of world situation. In Europe we now experience potential food insecurity directly through pricing and access to high quality food and diets.[2,3]

The productivity of the landscape in Northwestern Europe is one of the key reasons it is so difficult to convey the need for food awareness and security to family, friends, and colleagues. I have grown up in these landscapes that have had agricultural production supported and guided by government support and industrial investment. The situation is similar to the landscapes of Australia and the United States I have visited, where the support provided for a farming industry may be different but it still exists in some form. I know that this view may not be in line with everyone's thinking on the matter, particularly the issue of direct government support for farming. In the European Union (EU), the Common Agricultural Policy (CAP) is an important part of the European Commission (EC) spending, as is the Farm Bill in the United States.[4,5]

The Impact of Changing Worldviews

The idea of a working landscape that provides many functions beyond agricultural production has not always been a typical one. An incredibly influential piece of work for my students and myself of the time when Europe was considering the value of the whole food supply chain was written by Professor Jules Pretty in the late 1990s, titled *The Living Land*, which provided a view of changing agriculture in Europe.[6] These ideas were effectively packaged in the United Kingdom by the government's report 'Reconnecting the Food Chain' chaired by Lord Donald Curry of Kirkharle and known as 'The Curry Report'.[7] This time of change was one where we began to look at the whole food supply chain rather than thinking only of agricultural production or food manufacturing in isolation. In this environment of change, Jules Pretty's analysis developed the directions of many farmers I taught and convinced them to follow what became active and world-changing roles in world agriculture. The transition in agriculture at that time was one of coming from production-focused agriculture to one that included social and environmental value on equal footing with profitable agriculture.

Decoupling Production and Profit

The landscapes across Europe and the world have undergone a revolution in many ways during this period because of the need to consider the food supply chain beyond agricultural functions. This has important implications for the global food system that started with the first programmes aimed at curtailing agricultural overproduction in Europe appeared in the late 1980s. These were the so-called set-aside schemes, whereby farmers would place fields into a resting period or fallow voluntarily and receive a payment for doing so.[8] The focus of this set-aside scheme was to reduce agricultural production but it soon emerged that there were conservation and leisure benefits to set aside and they evolved into environmental stewardship schemes.[9,10] The schemes themselves became associated with different forms of integrated agriculture because it has been shown that enhancing biodiversity on farms is associated with pest management strategies, and maintaining soil fertility was associated with improved plant nutrition and water management. In short, a range of benefits became apparent, and this approach not only changed farming practice in a time when agricultural produce prices and farm input prices for fertiliser and feed were depressed but also changed how farmers and producers thought about their businesses. There was a transition from a primarily food production mindset or rather one of producing food ingredients to one of being part of a rural landscape. This meant integrating social awareness in terms of fairness, ethical production, tourism, and leisure within farming and food businesses.

Food is made more affordable with a productive agricultural system. The agricultural framework for this does not develop by increasing yield and quality attributes alone. The Living Land laid the ground for this, and in many ways, Professor Pretty's vision for 1990s agriculture in the United Kingdom was proven correct.

Wider Changes in the Food System

Individuals have had incredible impact in bringing the principles of general sustainability to policy development. An example is John Elkington, who has transformed how large organizations think about sustainability through the triple bottom line approach described in his book, *Cannibals with Forks*, which I first read in 1999.[11] The triple bottom line approach put forward by Elkington has transformed how senior officers of companies view future business with regard to value and values across the social agenda.[12] This is becoming evident with a 2010 Accenture survey of 766 chief executives worldwide: 93% see sustainability as important for the future of their businesses; 88% accept that they must drive new requirements through their supply chains; and 81% say they have already integrated sustainability into their businesses.[13] These surveys are now commonplace, and people like Elkington have helped the global arena define the sustainability problems facing them and what they need to do about them. It is not surprising that people like Elkington, who came from an activist background in environmentalism, are now asking for accelerated change.

Activists do not like to wait or see ineffective actions that achieve very little; they are disruptive. This is a major criticism that is currently facing international organizations that have developed major events, such as the Rio Conferences, and international protocols that are not without successes. These include the Montreal Protocol to reduce halogenated refrigerants, identified as the root cause of the growing ozone hole observed at the South Pole in the 1980s.[14] The Basel Convention developed international standards for the trade in waste materials, providing further requirements for businesses to act responsibly with regard to polluting impacts.[15] The Intergovernmental Panel on Climate Change (IPCC) of the 1990s established the requirement to define the impact of greenhouse gas (GHG) emissions, and this now packages several aspects of how businesses procure and utilize resources.[16] These actions have defined the issues we now face extremely well, and this book presents solutions to many of the issues they identify that are being delivered to the food arena by companies and their supply chains.

It is clear that materials and foods are becoming scarcer, and many are considered critical resources. The way in which we measure reserves of materials is always debatable, but organizations are taking actions that will deal with scenarios of critical resource availability. For example, the concept of 'peak resources' is well established for oil and phosphorus resources, and the peak scenarios are clearly dependant on exploration for new reserves; calculations typically consider the reserves of 30 years into the future.[17] This results in revitalised prospecting and exploration for resources each 30 years. This situation exists for most resources and should be borne in mind for our analyses here. For example, the impact of new technologies and new management systems can improve efficiencies of use or find new routes to conserve wastes.

What has become apparent for critical resources is that the quality of material found during exploration has decreased. This is the case for iron ore, for example, where a 1% decrease in iron ore results in more energy required extracting steel and significant costs in recovering ore. Together with peak scenarios, there are useful analogies we can make with food supply because similar scenarios are seen for high-quality land, which is ultimately the primary resource for food production. There is clear evidence of land trading or 'land grabbing' activities for biomass, biofuel, and agricultural food production since the 1990s, and the ability for land to provide efficient nutrient balance or protein production is under increasing stress. Resources that are currently considered to be at points of criticality in the global food system include water and phosphorus.[18,19] Water supplies are most stressed in parts of the world where crop production is or is likely to be most important, and they are most susceptible to the impact of temperature and precipitation changes that are the outcome of long-term climate change.

Resource regeneration and closed-loop thinking are most definitely subjects that are dominating the mining industry where the use of metals in consumer goods has identified both its criticalities and opportunities. The scenario for the critical metals arena is important to us because there are important cross-considerations for food supply globally; the issues of 'resource nationalism' that have become apparent are also emerging in the food system with water and land resources. The requirement to consider using materials such as protein more efficiently is as apparent as the global metals system, but in order to do this, new business models, such as those that Elkington and others have established, are needed. For example, in the case of metal supply, there is more gold in one cubic metre of mobile phones that are disposed of at the end of their product life than there is in a cubic metre of gold ore currently mined.[20] The problem faced by the electronic consumer goods industry is being able to recover gold and other metals, particularly rare earth metals, efficiently and cost-effectively. The possibility to generate geoeconomic conflict has been seen by restricting the trade of rare earth metals by China in the period 2005–2009, where prices of rare earths increased dramatically.[21] This response is termed 'resource nationalism', but it is something that has existed for centuries and it ultimately forces industries to consider new relationships, methods, and materials. In short, resource criticality can be debated in terms of geopolitical and geoeconomic factors, but it is stimulating innovation that aims to overcome current limits to supply; the same is true for all natural resources, and this type of critical thinking has important impacts on food supply chains.

The Food System

In the food system, we have seen criticality expressed by companies and their supply chain through the prospecting and use of phosphorus, water, and land. However, the food system itself is somewhat different from the other primary industry of mining for materials in that there are fewer opportunities for regenerating stock in supply chains because foods are ultimately digested. Of course, the recycling of minerals and nutrients are integral to any food production system and were perhaps the first recycling industries as identified by Lawes and Gilbert of Rothamsted, who are discussed later. Furthermore, the use of manufacturing and retailing infrastructures associated with a sustainable food supply chain are largely transferrable globally because of efficient logistics. Establishing sustainability in the food system does depend on trade and trade routes, and these are experiencing huge change globally, particularly in response to establishing trade in the Indian Ocean and the rise of the middle class associated with the growth of China and other emergent and emerging economies.

The Future of Food

Thus, this book places the importance of obtaining a supply chain approach in tackling food security and sustainable food supply where technical and social factors are integrated to provide solutions. The technical breakthroughs that will provide novel nutrition, safety, and design attributes of products will need to be integrated with consumption trends and a very clear understanding of how consumers taste and experience foods. For example, the use of genomics technologies that will provide libraries of materials to work with must be developed with a very good understanding of how consumers use and consume food products. This requires the capacity to assess increased amounts of information, and the need to use methods that visualize and provide a framework for information to be applied to security challenges will be increasingly important. In this book, the use of geographical methods and life cycle assessment (LCA) approaches are put forward as a means to help companies deal with the issue of analysing large and complex datasets in their supply chains.

References

1 Gandhi, I. (1975). *A world without want. Encyclopaedia Britannica* Book of the Year 1975 (pp. 6–17). Chicago: Encyclopaedia Britannica Inc.

2 Independent Commission on International Development Issues, & Brandt, W. (1983). *Common crisis North-South: cooperation for world recovery.* Cambridge, MA: MIT Press.

3 World Commission on Environment and Development (1987). *Our common future.* Oxford: Oxford University Press.

4 Lowe, P., Buller, H., & Ward, N. (2002). Setting the next agenda? British and French approaches to the second pillar of the Common Agricultural Policy. *Journal of Rural Studies, 18*(1), 1–17.

5 Moyer, W., & Josling, T. (2002). *Agricultural policy reform: politics and process in the EU and US in the 1990s.* Aldershot: Ashgate Publishing Ltd.

6 Pretty, J. (1999). *The living land: agriculture, food and community regeneration in the 21st century.* London: Earthscan.

7 Curry, D. (2002). Farming and food: a sustainable future (the Curry Report). London: Report of the Policy Commission on the Future of Farming and Food, UK Cabinet Office.

8 Bignal, E. M. (1998). Using an ecological understanding of farmland to reconcile nature conservation

requirements, EU agriculture policy and world trade agreements. *Journal of Applied Ecology*, *35*(6), 949–954.

9　Robinson, R. A., & Sutherland, W. J. (2002). Post-war changes in arable farming and biodiversity in Great Britain. *Journal of applied Ecology*, *39*(1), 157–176.

10　Sutherland, W. J. (2002). Restoring a sustainable countryside. *Trends in Ecology & Evolution*, *17*(3), 148–150.

11　Elkington, J. (1997). *Cannibals with forks*. Oxford: Capstone.

12　Elkington, J. (1998). Partnerships from cannibals with forks: the triple bottom line of 21st-century business. *Environmental Quality Management*, *8*(1), 37–51.

13　Elkington, J. (2010). Agenda for a sustainable future. Cheng, W., & Mohamed, S. (Eds.). (2010). *The World that Changes the World: How Philanthropy, Innovation, and Entrepreneurship Are Transforming the Social Ecosystem*, pp 359–371. John Wiley & Sons.

14　Velders, G. J., Andersen, S. O., Daniel, J. S., Fahey, D. W., & McFarland, M. (2007). The importance of the Montreal Protocol in protecting climate. *Proceedings of the National Academy of Sciences*, *104*(12), 4814–4819.

15　Kummer, K. (1992). The international regulation of transboundary traffic in hazardous wastes: the 1989 Basel Convention. *The International and Comparative Law Quarterly*, *41*(03), 530–562.

16　Burton, I., Huq, S., Lim, B., Pilifosova, O., & Schipper, E. L. (2002). From impacts assessment to adaptation priorities: the shaping of adaptation policy. *Climate policy*, *2*(2), 145–159.

17　Bardi, U. (2009). Peak oil: the four stages of a new idea. *Energy*, *34*(3), 323–326.

18　Vörösmarty, C. J., Green, P., Salisbury, J., & Lammers, R. B. (2000). Global water resources: vulnerability from climate change and population growth. *Science*, *289*(5477), 284.

19　Cordell, D., Drangert, J. O., & White, S. (2009). The story of phosphorus: global food security and food for thought. *Global environmental change*, *19*(2), 292–305.

20 Reller, A., Bublies, T., Staudinger, T., Oswald, I., Meisharp-ner, S., & Allen, M. (2009). The mobile phone: powerful communicator and potential metal dissipator. *GAIA-Ecological Perspectives for Science and Society, 18*(2), 127–135.

21 Du, X., & Graedel, T. E. (2011). Global in-use stocks of the rare earth elements: a first estimate. *Environmental Science & Technology, 45*(9), 4096–4101.

1 The Basis for Food Security

1.1 Defining What Food Security Is and How Food Supply Chains Can Deliver It

This chapter will clarify many of the complex definitions of food security so that we can relate them to the food supply chain and food system. Our analysis will present the findings from research, agricultural field trials, and industrial case studies that have shaped the current food system. As previously described, food security is often a difficult attribute to describe adequately because it is the sum of many aspects of our lives. Food security is concerned not only with the immediate supply of protein and energy, but also the sustainable supply of a healthy diet that promotes well-being. While the immediate requirement for protein and energy is critical, security will also include what we experience as accessibility, affordability, and availability of foods when we consume them as meals and our diet.[1] Indeed, an important viewpoint put forward in this book is that food security should consider all of us as consumers rather than recipients of food. While this is a descriptive point, it is important to understand that food security is not only an issue of quantity, it has become an issue that is increasingly identified by quality of life and safety attributes that are delivered to consumers.

Global Food Security and Supply, First Edition. Wayne Martindale.
© 2015 John Wiley & Sons, Ltd. Published 2015 by John Wiley & Sons, Ltd.

In such a context, the food supply chain provides all the criteria necessary for food security, and this means the components of the supply chain must operate efficiently. Defining the components of supply and consumption is an important first step in understanding food supply chain efficiency. The food supply chain operations that make food security a possible goal are dependent on the production of ingredients and raw materials from agricultural operations and the development of food products by manufacturers and processors. A critical function of the food supply chain that is extremely variable and the focus of much attention because of the relationship to consumers is that of the distributor, wholesaler and retailer, who make sure that food is presented to the consumer. Thus, these agricultural, manufacturing, retailing and consumption aspects of the food supply chain can be presented as a series of four functions that are shown in Figure 1.1. The food supply chain functions will be discussed and investigated in further chapters, but the role of Figure 1.1 is to explain food supply with elements of simplicity that are the key to us developing ideas in the further chapters of this book.

Naturally, the simplicity presented here is fine for explanation of principles, but when this supply chain model is applied to populations, it becomes very complex due to several other attributes associated with the impacts, services, and capital of businesses and consumers that require consideration. The supply chain shown is easy to understand, but projecting it to populations and millions of consumers means it becomes potentially impossible to visualise. The sheer scale of supply functions in populations and the variance of inputs and outputs into food supply chains globally result in the need to consider the model presented in Figure 1.1 as a food system. Scientific and sociological research has provided evidence that shows how the development of food supply chains can result in the establishing of an understanding of what makes a food system sustainable.[2] These ideas will be developed, but an

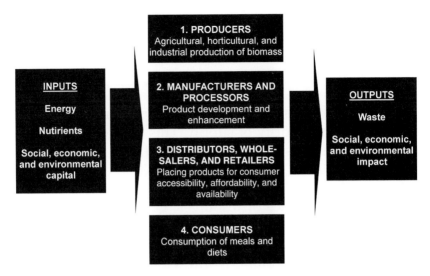

Figure 1.1. The food supply chain functions and food system. There are four functions to the supply chain scenario presented here: producers; manufacturers and processors, distributors, wholesalers, and retailers; and consumers. Inputs and outputs can be measured as a balance or LCA function at each function. This is a relatively simple model, but it becomes complex when applied to populations and several supply chains.

understanding of different types of inputs and outputs from the food system is central to the ideas put forward and critical in determining our perspective on food security.

An important consideration for each part of a food supply chain as a producer; manufacturer and processor; retailer, distributor, and wholesaler; and a consumer is to consider where products are being made and where they are being used. Understanding these two parts of the supply chains is critical, and it has been traditionally defined by supply and demand functions that determine what consumer trends are evident. This view of supply chains has been established for centuries and as we will see it has now developed to consider other value aspects of goods at the end of the twenty-first century. These values are associated with social and environmental impacts, as well as economic wealth creation, and they have increasingly become coupled with criticality of supply for specific resources. That is, the supply and demand functions of supply chains must increasingly be affiliated with assessment of security of supply. This is true for most manufactured goods, it always has been, but new pressures have emerged to make an understanding of the 'push or pull' components in supply chains. These include rapidly changing abundance of resources, geopolitical structures, price variability, price volatility, environmental impact, and health impact. Assessments of these attributes together can provide important world views on the risk of limited supply when they are blended with the consumer trends. Food supply chains provide the data concerned with material flows that enable the assessment of risk and uncertainty of ingredient and food supply. Thus, understanding where foods come from and where they are used allow us to project trends in consumption and allow us to develop strategies that deliver resilience in response to volatility and geo-political change.

Identifying the attributes of supply chains that can determine trends and criticality of supply are well characterised and have been for a significant period of time now. For example, the thought-provoking 'Limits to Growth' reports identified population growth, availability of natural resources, pollution, and capital investment in food supply chains as critical points in delivering sustainable global food supply.[3] Whereas security assessment of supply chains is well developed for minerals and metals, it is perhaps less so for food products. The key players in providing this assessment of security are those involved in the supply chain functions, that is, the producers, manufacturers, retailers, and consumers. Understanding price variability and volatility of resources is crucial to developing trends and strategies for dealing with risk and uncertainty associated with food supply. The time scales that are used can change our perspective on sustainability because many assessments will consider data from a time series of 5 years even though we might consider projections of decades into the future most important. Price data can be used for longer periods, and we should always consider the value of using longer term time series that are greater than a 5-year historical record.[4] It is notable to observe the recent price spikes in food globally that augmented the current food security debate, and the value of using 5-, 10-, 20-, 50-, or 100-year historical price series will provide different projections for security.[5] Thus, a consideration of the attributes we use to develop trends is just as critical as the time series we utilise to develop food security projections.[6] The caveat placed by this study and book is that this cannot be done without considering the food supply chain due to the supply chain being both the provider for trend data and a source of innovation that enables the delivery of food products that consumers demand.

Indeed, the need for organisations and businesses to rank the materials they utilise in terms of the risk associated with supply has become more important since the food price spikes of 2005–2008, otherwise called the 'perfect storm' scenario.[7] The perfect storm was a convergence of increased demand for livestock products and a diversification of agricultural biomass into liquid biofuels. This created price volatility and uncertain supply. In a similar way, the trade of steel used for industrial infrastructure and rare earth metals used in electronic goods experienced extreme price volatility at the same time. The price hikes have focused our thinking around security with regard to our considerations of sustainability and the amount of resource reserves that are available to food supply chains.[8] That is, how much genetic biodiversity, useful land, minerals, metals, and fossil fuels are available to produce food products. The current security debate has not only considered quantity of food, but it has begun to consider the quality attributes of food with nutrition that delivers well-being. That is, reductions in quality blended with increased production of biomass, ingredients, and food products may not provide the benefits we think because of the impact on energy consumption during manufacture and health impact after consumption.

The consideration of closed loop economies has emerged from security crises, these are systems that eradicate or reduce waste from supply chains so that everything used to produce a product remains within the supply chain or linked with other supply chains. Closed-loop thinking is different for food products because it is usually consumed and cannot be re-eaten. However, food waste within supply chains is of critical importance to future security and sustainable supply. Furthermore, nutrients manufactured into food products can be recycled within the food system to support the production of biomass. Indeed, the production of composts and manures for agricultural systems is likely to have been our first experience of recycling materials several thousand years ago. We do increasingly know more detail about the environmental and social impacts of food products due to increased access to data that are either open sourced or peer reviewed. These show the emission factors and mass-flows for food ingredients and products. Thus, for the first time, we can now identify criticality points in food supply because of the economic impetus to do so. This can be integrated with measures of sustainability for the first time historically, and it is being done by food supply chains that will survive the 'perfect storms' of the future.

1.2 The Convergence of Food Security Research, Economics, and Policy

Specific analytical methods are often employed to measure inputs and assess the impact of outputs from the food system that traditionally identified economic and mass flows through the food system. This approach used to overlook social and environmental services, and this proved limiting for anyone who required a measure of current and future performance of the food system. Therefore, if we are to project future food security and sustainability taking a purely economic view based on production of foods, it would be a very fragile representation of the food system. The limits to such economic assessments were explored in a 1997 *Nature* paper by Professor Robert Costanza and colleagues, who estimated the ecosystem service worth of the globe to be two to three times that of the economic wealth.[9] This paper changed the way we think about natural systems and the sustainability of the global food system; it has also extended our views on how food security could be delivered. The global policy-making environment established by the World Commission on Environment and Development report 'Our Common Future' 10 years before demonstrated distinct convergence of views from policy and research on sustainable natural resources.[10] The 'Our Common Future' report established the United Nations (UN) Conference for Environment and Development or 'Earth Summit' and a set of targets for the new millennium known as Agenda 21. The paper by Costanza and colleagues provided an assessment of integrating the goals of sustainable development with the financial risk if we were not to meet sustainability targets that were increasingly being developed by policy-makers, and understanding this risk was critical to the future of humankind. The Costanza and colleagues' *Nature* paper essentially stated that we should be very aware that inaction on sustainable development could be associated with trillions of dollars of risk and gave the following description in its opening statement:

The services of ecological systems and the natural capital stocks that produce them are critical to the functioning of

> *the Earth's life-support system. They contribute to human welfare, both directly and indirectly, and therefore represent part of the total economic value of the planet.*

This 1997 paper changed how we measured and assessed the food system, and it also related inaction on food security to very clear financial risks. This was important because the Costanza paper demonstrated that the growth of economic capital was clearly influenced by both social and natural capital.

1.3   The Millennium Development Goals (MDGs)

The convergence of research and policy research in food security resulted in the bold establishment of the Millennium Development Goals (MDGs), which is an international programme developed from approaches established in the 'Our Common Future' report in 1987. While the international landscape of agreement, conference, and commissions can seem unpractical, the dialogue they have established has changed from one of describing problems to one of achieving and meeting specific goals, which are shown in Figure 1.2.

The first MDG aims 'to eradicate poverty', and it uses assessments of economic, social, and natural capital to identify how food insecurity can be alleviated in the world. The progress to the target is recorded, and this achieves an establishment of accountability. Thus, the MDG's do provide the important starting point for our study of food security and supply utilising the ecosystem service approach. We are approaching the point where we know whether we can achieve the eight MDGs laid down for the global community in September 2000 during the Millennium Summit of the UN. This was one of the largest ever gatherings of heads of state, and they marked the new millennium by adopting the UN Millennium Declaration.[11] This was endorsed by 189 countries, and it established a

1. Eradicating extreme poverty and hunger

2. Achieving universal primary education
3. Promoting gender equality and empowering women

4. Reducing child mortality rates
5. Improving maternal health
6. Combating HIV/AIDS, malaria, and other diseases

7. Ensuring environmental sustainability
8. Developing a global partnership for development

Figure 1.2.   The MDGs as described by the UN; the MDG 1, 4, 5, and 6 relate directly to issues of diet and sustainable nutrition for well-being.
Source: Adapted from *Road map toward the implementation of the United Nations Millennium Declaration*. New York: United Nations, 2002. United Nations General Assembly Document A56/326.

roadmap for achieving the goals to be reached by 2015.[12] Roadmaps have become important illustrative tools for policy development, which put forward agreed targets and suggested routes to obtaining them. The MDGs are important because they have provided targets and a new round of questioning the actual competencies of international efforts.[13]

The MDGs are important as they provide a standardised measure of progress towards food security that have been agreed by the members of the UN. These both provide a means to compare progress across nations, and it gives a form of consensus on what is required to provide food security. The MDG Target of halving those people experiencing hunger is reported as 'within reach' in the UN MDG Report 2013[14] An important consideration of the MDGs is the spatial variation in attaining them, with sub-Saharan Africa and South Asia presenting the most acute areas for food security concern by being most at risk.

The UN Food and Agriculture Organisation (FAO) reports progress in reaching the MDGs, and they have stated the following in 2013 for achieving an MDG1 Target of halving the number undernourished people since 1990:[15]

The [revised results imply that the] Millennium Development Goal (MDG) target of halving the prevalence of undernourishment in the developing world by 2015 is within reach, if appropriate actions are taken to reverse the slowdown since 2007/08.

This Target is one of 21, and it is part of the first Goal, and it is fraught with controversy because of the period between 2005 and 2008, which saw a new phase of development in the global food system that exposed fragility in supply. This was due to the demand for agricultural products from national economies that had globalised since the 1990s and most notably resulted in food price increases in 2010 that had significant impact on the world food system.[16] Changes in food price and affordability have severe impacts on the number of people experiencing hunger, and the price spikes of 2010 resulted in hundreds of millions of people experience poverty or extreme poverty.[17] Globalisation of the food supply chain shown in Figure 1.1, produced changes in consumption, tastes and society that had dramatic impact of where agricultural commodities were traded and what they used for.[18,19] As such, the MDGs begin to describe the complexity of what food security is because they not only highlight the supply of resources to eradicate poverty and hunger, but also consider access to, safety of and education regarding the use of natural resources. The 2012 FAO report, 'The State of Food Insecurity in the World', which is published annually states the following.[20]

The State of Food Insecurity in the World 2012 presents new estimates of the number and proportion of undernourished people going back to 1990, defined in terms of the distribution of dietary energy supply. With almost 870 million people chronically undernourished in 2010–12, the number of hungry people in the world remains unacceptably high.

The FAO reported that the vast majority of these people live in developing countries, where about 850 million people, or in some cases close to 15% of the individual nation state populations, are estimated to be undernourished.

1.4 Measuring Hunger in a Changing World to Establish Security

The Global Hunger Index (GHI) provides a descriptor of how we assess hunger, and it is published by the International Food Policy Research Institute (IFPRI), Concern Worldwide, and Welthungerhilfe.[21] The 2012 GHI shows that progress in reducing the proportion of hungry people in the world is slow and hunger on a global scale remains 'serious'. Twenty countries still have levels of hunger that are 'alarming' or 'extremely alarming'. South Asia and sub-Saharan Africa continue to have the highest levels of hunger. The 2012 GHI is the seventh year that IFPRI has calculated it, and the country averages represent variable data, so those countries classified as having 'moderate' or 'serious' hunger can have specific areas where the situation is 'alarming' or 'extremely alarming'. This has important implications for mapping food supply and security because the use of maps can effectively convey and describe variability in large data sets that include the social, cultural, and ecosystem attributes used to describe hunger.[22,23]

The development of the GHI and other measures associated with food supply and security that take into account spatial variation within regions is likely to be an important future development. The resolution of food security data spatially is an important component of future food policy and the deployment of actions that tackle food security. Naturally, population census data and the reporting of food supply statistics are critical to any measure of spatial variation, and the coverage of agricultural, food industry, and population census globally will limit these actions.[24,25]

Spatial variation of food security is not the only reason why measuring hunger is not straightforward. It can be described in many ways; as with food security, it is multi-dimensional. The GHI combines three equally weighted indicators that are broadly agreed upon across many organisations dealing with it and combines them as a single index. The three indicators that are combined are now described:

1. **Undernourishment:** The proportion of undernour-ished people as a percentage of the population (reflect-ing the share of the population with insufficient caloric intake).
2. **Child underweight:** The proportion of children younger than age five who are underweight (i.e., have low weight for their age, reflecting wasting, stunted growth, or both), which is one indicator of child undernutrition.
3. **Child mortality:** The mortality rate of children younger than age five (partially reflecting the fatality of inade-quate caloric intake and unhealthy environments).

The GHI aims to provide insight into the nutrition situa-tion of not only the population as a whole, but also chil-dren who are a physiologically vulnerable group where a lack of nutrients leads to a high risk of illness, poor physi-cal and cognitive development, and death. The GHI ranks countries on a 100-point scale in which zero is the best score (no hunger) and 100 the worst, although neither of these extremes is reached in practice.

'Hunger' is understood to refer to the discomfort associ-ated with lack of food and the FAO defines food deprivation or 'undernourishment' specifically as the consumption of fewer than 1,800 kcal a day. This is determined to be the minimum that most people require to live a healthy and productive life and anything under this would be considered undernutrition. The FAO considers the composition of a population by age and sex to calculate its average minimum energy requirement, which varies by country.

The FAO defines food insecurity as the following.

A situation that exists when people lack secure access to sufficient amounts of safe and nutritious food for normal growth and development and an active and healthy life. It may be caused by the unavailability of food, insufficient purchasing power, inappropriate distribution, or inadequate use of food at the household level. Food insecurity, poor conditions of health and sanitation, and inappropriate care and feeding practices are the major causes of poor nutritional status. Food insecurity may be chronic, seasonal or transitory.

Thus, even undernutrition is more than a consideration of the consumption of calories because it includes deficiencies in energy, protein, or essential vitamins and minerals. Undernutrition can be the result of inadequate intake of the quality or quantity of food and includes poor bioavailabity of nutrients because of disease or reduced quality of food consumed. Overnutrition represents the problems of unbalanced diets that are largely caused by the consumption of too many calories and are often associated with poor micronutrient quality of diets.

1.5   The Undernutrition and Overnutrition Gap

Figure 1.3, demonstrates the gap between undernutrition and overnutrition globally. The FAO data shown are derived from national agricultural and food industry

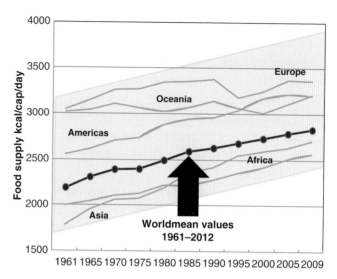

Figure 1.3. The undernutrition gap demonstrated for global mean calorific supply (FAOSTAT data).
Source: **These data were adapted from FAO. FAOSTAT (2009). Food supply, crops, primary equivalent data set. http://faostat .fao.org/ (accessed 22 April 2014).**

census. The figure shows the average supply of calories to individuals has increased year on year since 1961, but there have been clear inequalities in how supply is distributed globally. Asian and African regions have consistently been below this average and are cited in IFPRI and FAO reports that measure undernourishment as areas of most concern. European, American, and Oceanic (principally Australasian in this instance) are Calorie sufficient.

The Brandt report defined these spatial inequalities and it was published in 1980 by an independent commission Chaired by Willy Brandt, chancellor of Germany from 1969 to 1974.[26] At its most basic, it provided an understanding of drastic differences in the economic development for both the North and South hemispheres of the world. The Brandt report is important because it raised the issue of the 'standard of living' differences that exist along the global North–South divide, and there should therefore be a large transfer of supporting resources from developed to developing countries. It most certainly did not foresee the extent of globalisation and the emergence of the impact large economies, such as China, would have on the world food system. In many respects, it presents the prior world view of globalisation and even though the inequalities it identified do still exist, as shown in Figure 1.3, the impact of globalisation was not foreseen. Indeed, the type of analysis led by Brandt has been reanalysed and perhaps most emphasised by Professor Jared Diamond in his book *Guns, Germs and Steel*.[27] These analyses firmly place the role of limitations placed on the food system by land and climate to be considered in delivering food security because they will influence the capacity to supply food. While technologies will alleviate these limitations, the access to them may be again limited or controlled by trade.

1.6   The Supply Chain and Nutrition Gaps

We can increasingly see that delivering food security is not only a case of producing ingredients and foods, there is an absolute requirement for highly efficient supply chains that deliver safe and nutritious food to consumers. While supply chain functions can be viewed simply as a series of four components shown in Figure 1.1, they quickly become complex at the scale of populations. Complexity in supply becomes apparent when the number of consumers, suppliers and ingredients used increase. Figure 1.3, demonstrates the impact of these principles on current global calorie supply. The world food supply of protein actually approaches 80 g for each person per day, as reported by FAO statistics in 2009, which is considered to be sufficient for most of the global population, at least a third of this can be from cereals.[28] Advice from FAO and WHO state 0.66 g of protein per kg of body weight is sufficient for a healthy diet that maintains health.

The FAO and WHO figure now combine a 1973 protein recommendation that relied on a limited number of short-term and longer-term nitrogen balance studies with a later 1985 study to derive the protein requirement of adults. Some of these studies were designed to identify a requirement and others to test the safe level (0.58 g protein/kg per day). Taken together in the 1985 report, they were interpreted as indicating a mean requirement of 0.6 g protein/kg per day, with a coefficient of variation estimated to be 12.5% in a typical population. This resulted in a safe protein intake recommendation of 0.75 g/kg per day; that is, a value at 2SD above the average requirement (0.66 g protein/kg per day), which would provide for the needs of nearly all individuals (97.5%) within a target population.

Human metabolism regulates protein consumption to about 15% of total intake of a meal, and this is a much stronger relationship than for fat and carbohydrate.[29] This will be returned to later, but there is strong evidence to suggest that the sustainability of healthy diets can be measured by protein content. Indeed, protein intake leverages itself against fat and sugar consumption in that when protein intake is low, consumption of fat and sugar may increase in order to reach the 15% level of protein intake.[30] As previously discussed, the average global protein supply hides large variations, with the protein supplied to an average European each day being 102 g, some 35 g more than a typical African. This variation describes the range of protein deficiency and oversufficiency, and it can also hide significant changes in the protein balance of diets shown in Figure 1.4. The protein supplied from cereal crops in Europe has declined since 1961, whereas during the same period, it has increased for the Asian region, where there are severe food security impacts, as highlighted by the GHI assessment. This transition identifies how regions can increase protein supply and change protein consumption dynamics globally. The average Asian

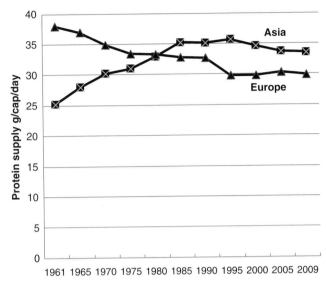

Figure 1.4. Protein supply from cereal crops in Asia and Europe1961–2010.
Source: **These data were adapted from FAO. FAOSTAT (2009). Food supply, crops, primary equivalent data set. http://faostat .fao.org/ (accessed 22 April 2014).**

citizen will be supplied with some 30 g less protein than a European citizen on a daily basis, but more of the protein in an Asian diet is from cereals than a European one.

This is important because dependency on cereals for balanced protein and nutrition depends on efficient production, and it has provided an important target for improving health. For example, the development of golden rice varieties with enhanced carotene content has been developed by large agricultural companies that have included Ciba Geigy and Monsanto and the resulting golden rice Humanitarian Board.[31,32] The controversy surrounding the genetic modification technologies used to produce these varieties has met with resistance even though there is a clear nutritional benefit to consuming biofortified crops, such as golden rice. There are very clear indications that the pro-vitamin A activity of enhanced carotenoids also improve the bioavailability of other micro-nutrients, such as iron.[33]

1.7 The Relationship between Food Security and Biology

What has changed during the course of progress to the MDG Target for halving hunger is that the focus for tackling food security has become far broader than tackling agricultural limitations. Even though it is clear that improving agricultural production does alleviate pressure on the food system with respect to supplying a nutritious diet, there are many other components of the food supply chain that will restrict access to food if they are not considered. It has become clear that the requirements for economic growth and social protection to be embedded in policies that are sensitive to supplying a nutritious diet are critical in delivering food security.[7]

An understanding of how sciences have influenced changes in food security policy is as necessary as appreciating the role of social, political, and economic change. With this in mind, the application of bioscience to agriculture has improved food security for billions of people, and it was the focus of alleviating hunger in the twentieth century. The Green Revolution of the 1960s and 1970s saw crop yields per hectare rise dramatically because of the application of crop breeding programmes that were targeted for specific global regions. Reassessments of cropping and grazing management globally provided a way of identifying where crop breeding, engineering, and management programmes could be applied to increase the yield of biomass per hectare of land. These approaches at the time were effectively led internationally by Dr Norman Borlaug, who was awarded the Nobel Peace Prize in 1970 and established the World Food Prize in 1987 for research excellence applied to the whole food system.

This was a time when it was clear what needed to be done because the risks of not doing anything to alleviate food insecurity were being made all too clear by research presenting the limits of natural resources, such as that of Professor Paul Ehrlich's population time bomb ideas put forward at the end of the 1960s.[34] Dr Norman Borlaug recognised the potential for tailoring crop varieties for specific global regions and particularly arid regions where water limitation and hunger were likely to be chronic problems. The agricultural system was critical to alleviating the threat of famine that Ehrlich's book laid out to the world, and Borlaug offered an option for overcoming the limits imposed by the agricultural systems of that time. By developing crop varieties that would enable crop yield to remove the scourge of hunger from the lives of millions of people, he demonstrated field agronomy had a critical role to play. He stated the following:

Civilization as it is known today could not have evolved, nor can it survive, without an adequate food supply.

Borlaug's approach to alleviating the limits of food supply began in 1944, when he participated in the Rockefeller Foundation's wheat improvement programme in Mexico as a research scientist working on wheat production problems that were limiting wheat cultivation. This developed disease- and climate-resistant varieties, but also trained scientists to develop new methods of managing crops by a process known as agricultural extension. The field basis for extending agricultural research to production has been critical to developing agricultural systems that can alleviate hunger. New wheat varieties with improved yield potential and improved crop management practices transformed agricultural production in Mexico during the 1940s and 1950s and later in Asia and Latin America, sparking what today is known as the 'Green Revolution'. This approach has led to science and crop breeding saving millions of lives, with the World Food Prize Foundation stating the following, which demonstrates the value of extending the actions taken by crop scientists to the saving of human life by alleviating hunger.

Because of his [Dr Norman Borlaug's] achievements to prevent hunger, famine and misery around the world, it is said that Dr. Borlaug has "saved more lives than any other person who has ever lived."[35]

What Borlaug did was to show a clear way of alleviating food supply limits by improving agricultural productivity. The rule base for doing this was relatively simple in that agriculture should be fit-for-purpose in specific environments that have different climate, soils and needs. If we consider the words of Dr Daniel Hillel, who was awarded the 2012 World Food Prize, in his book *Out of the Earth*, it is clear that these things often overlooked:

All terrestrial life depends on soil and water. So commonplace and seemingly abundant are these elements that we tend to treat them contemptuously.[36]

The World Food Prize does provide an indicator of change with regard to tackling many of the challenges the food system places before us. Tackling the food insecurity that has limited the ability of humankind to ease conflict and the unsustainable use of natural resources entails using multidisciplinary approaches. The World Food Prize reflects this with the Laureates being from the policy, agronomy, biotechnology, and human health arenas.[35] Development of new crop varieties and methods of growing crops feature strongly in the list of Laureates, along with those that are focused on the policy and political development of efficient supply chains. This demonstrates the duality of achieving food security in that the limitation of the supply function can be alleviated by efficient agricultural production that taps into genetic resources of crops and livestock. However, security can be achieved only if social and political will support supply functions from the farm to the consumer. What has become apparent with the World Food Prize is the need to reevaluate what Borlaug achieved by improving farm efficiencies because it is clear that production and yield of crop varieties and livestock breeds have improved globally, so what do we do now to maintain efficiency?

This is apparent if we consider yield improvement of major crops since the 1960s, where yield increases between 2% and 10% year-on-year have been observed for crops that are crucial to the production of ingredients and feed. These types of data set also raise extremely important considerations of weather and climate because from year to year, changes in temperature and water availability have significant impacts. The year-on-year yield increase of wheat grain between 1961 and 2012 for the United Kingdom and world has been 0.8% and 1.9%, respectively (FAOSTAT data). Figure 1.5a,b shows that these data sets hide variation in yield percentage change year-on-year during this 52-year period. Understanding this variation is important because it enables the implementation of structured adaptive management that can both deal with the uncertainty associated with weather and the trends associated with environment change. A notable recent study of this issue of long term changes and trends is that of Schlenker and Roberts (2009), who used long-term data sets to demonstrate that when the corn, soybean and cotton crops of the USA experience temperatures of above 29°C, yield decreases are seen and can be projected with confidence limits.[37]

Schlenker and Roberts (2009) suggest their research shows limited historical adaptation of seed varieties or management practices to warmer temperatures because they also accounted for changes in farm practice, such as water storage in response to increased temperature. They predict average yields to decrease by 30–46% before the

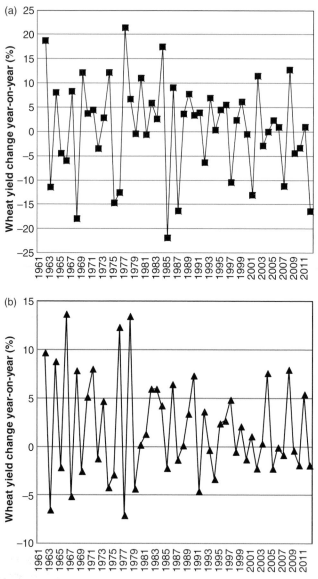

Figure 1.5.   (a) Changes in the yield of wheat grain year on year in the United Kingdom 1961–2012. (b) Changes in the yield of wheat grain year on year globally 1961–2012.
Source: These data were adapted from FAO. FAOSTAT (2012). Production, crops data set. http://faostat.fao.org/ (accessed 22 April 2014)

end of the century under the slowest global warming scenario and decrease by 63–82% under the most rapid warming scenario. Naturally, this has extreme implications for the food system and offers a target for plant breeders to select for crops that can tolerate extreme changes in temperature and water availability. Their research also highlights the requirement to work with long-term data sets when dealing with natural systems and the issues of food security. Working with long-term data should also implement adaptive management to account for changes in the environment and climate as they become apparent.[38] Adaptive modelling integrates well with the food system because records regarding biomass yield and production are often long term.[39] Indeed, the use of adaptive modelling is a key part of what we might consider integrated practices in agriculture and other parts of the food supply chain.

The recognition of the role crop breeding and agronomic technologies must have on yield improvements has enabled continued yield improvements, but there is a requirement for adaptive modelling and actions to be in place to account for environment change and global warming. Significant yield gaps do still remain if we take a global view of crop yields, and these are not closed by the development of technologies alone, they must be adaptive and respond to changes in environment, markets, and consumers.[40] The yield gap principle is presented in Figure 1.6 for wheat, rice, and maize, whereby the maximal national yield is compared with the global mean yield value of a crop and the resulting gap between them is considered the 'yield gap'.

As with most global and national production statistics, an understanding of variability in data must be emphasised with the fact that they will simplify complex supply chains. Regarding these cautions, the yield gap scenarios do still provide useful insights into what might be achievable and how maximum yields globally might be reached.

Cropping systems are the ultimate start of the food supply chain because photosynthetic biomass initially

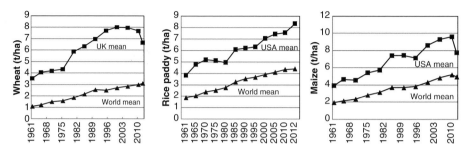

Figure 1.6.   Global yield gaps for wheat, rice, and maize. These yield gaps can be represented as the percentage gap between the greatest and world average yield of biomass. The percentage yield gap can be simply understood as an indicator for reaching the capacity for maximum yield of biomass. For example, the percentage yield gaps for wheat, maize, and rice shown are 53%, 36%, and 47% for the year 2012.
Source: **These data were adapted from FAO. FAOSTAT (2012). Production, crops data set. http://faostat.fao.org/ (accessed 22 April 2014).**

captures solar energy that is consumed by humans either directly or via livestock, or transformed via microbial fermentation onto food materials. Notable production systems based on microbial fermentations will be discussed in the following chapter because they offer potential to alleviating nutrition gaps globally, but Figure 1.6 describes crop yield gaps for three of the major food crops globally.

The limitation of defining production inefficiencies by crop yield gaps alone is fraught with difficulties because agriculture provides only the beginning of the food supply chain. The supply of food to consumers is dependant on all of the functions shown in our initial food supply chain model shown in Figure 1.1. For example, it is now apparent that the impact of waste and spoilage within supply chains is having a critical effect on getting food to consumers.[7,41] Furthermore, the role of processors and manufacturers in developing nutritious foods that are efficiently preserved does have a central role in the supply chain.[42] Indeed, the central place the processor and manufacturer occupies in supply chains between farmers producing ingredients and retailers who present food products to consumers is a critical one. Indeed, the case for adaptive modelling across food supply chains has not been used creatively and offers great opportunity to closing yield gaps and providing a more secure food system. The quantification of determinants of crop and livestock yield, such as different management practices, breeding programmes, and new technologies are rarely specifically defined in measures, such as Total Factor Productivity (TFP) and Life Cycle assessment (LCA). While TFP and LCA offer important methods for traceable analysis of products there is a requirement for assessment of respective components of crop and livestock yield that identify gaps in both the application of technology and the development of effective management that will help to close them.

This requires the assessment of both quantitative data concerned with the measurement of processes such as growing crops and qualitative data that describe how people feel about the implementation of new methods of production or development of new products. The application of biotechnology demonstrates the need to not only understand the process of developing solutions using technologies but also to engage users of technology, in this case consumers of food in the most appropriate way. The food industry as a whole understands this relationship because it delivers safe and wholesome food that has a very clear hedonistic purpose associated with it. That is, the food industry works with mixed research methods, both quantitative and qualitative at all times and this is well characterised. If we like how food tastes and performs when cooking and eating it, we are likely to consume it. Measuring this principle of hedonism is not as straightforward as it may seem, and it is explored with case studies and data later, but it is the basis for developing and managing sensory panels in the food sector. Sensory panels are groups of consumers who regularly test and assess food products;

the data from these panels are used to compare food products in terms of value, quality, taste, and performance.

1.8 The Relationship between Food Security and Biotechnology

The emergence of biotechnologies that can transfer specific genes between crops and livestock and thus transform traits that will improve production in the way that Borlaug did for cereals has revolutionised agriculture. The 2013 World Food Prize reflects this new biotechnological dialogue that accompanies the programmes that will tackle food insecurity in the world because it has been awarded to researchers whose work will impact on both the quantity and quality of biomass produced from farms. Dr Marc Van Montagu's research established the role of the *Agrobacterium* spp. plasmid stable vector for introducing genes into crop plants but this work crucially needed the involvement of crop technology companies. The 2013 Prize was co-awarded to Dr Mary-Dell Chilton and Dr Robert Fraley of the Syngenta and Monsanto group of companies. This addition to the Laureates helps us to define a change in reassessing Borlaug's legacy because the application of directed gene transfer and biotechnology is not only set to improve crop and livestock yield but improve nutritional quality.

An important example of quality improvement has been the establishment of agrifortification and biofortification; both result in the nutritional enhancement of biomass. Agrifortification achieves this by utilising mineral additions to food ingredients within the supply chain. Biofortification is a more stable process in that the nutritional enhancement is achieved by using the crop or livestock metabolism to improve nutritional aspects. This can be achieved by gene transfer, such as Agrobacterium-mediated transfer of genes or by traditional selection of varieties and breeds to achieve the desired response. Perhaps the most publicised example of biofortification mediated by gene transfer has been the development of the golden rice, but the principle of biofortification offers much hope in alleviating the so-called hidden hungers.[43] Golden rice was developed by the Ciba Geigy company to tackle a specific nutritional problem that has hampered the nutritional benefits of consuming cereals.[44] That is, cereals can provide a good source of energy and protein, but they are often less nutrient dense in the case of minor nutrients, such as minerals and vitamins. Vitamin A in rice was a significant challenge because it was principally consumed where agrifortification was not possible and cereals are a major component of diet. Vitamin A deficiency is known to increase the prevalence of a blindness associated with decreased consumption of foods containing the precursors of vitamin A.[45] Golden rice varieties are biofortified using genes that enabled rice seeds to accumulate vitamin A precursors and thus alleviate deficiency in diets where rice is the major food staple.[46]

An important aspect of developing new crop varieties is to understand how consumers cook and utilise them so that we can be sure that they meet the sensory and textural targets of being included into recipes and meals. This aspect of new crop variety development has been often overlooked, and the case study that golden rice provides is of critical importance to the market acceptance by consumers.[47] While elucidating the attributes associated with consumer use and market entry have traditionally been the interest of commercial functions of the supply chain, they are increasingly important to all aspects of supply and development, including the development of initial research and development concepts for new crop varieties. The concern consumers have about genetically engineered crop varieties has been established since the late 1990s, and recent food security issues have alleviated many of these concerns by considering the safety of products and their role in the whole supply chain; this includes consumer use and acceptance.[48] The development of consumer panel data in these types of situation where new ingredients and products are being developed has an important role to play that is likely to be more apparent in the future development of genetically engineered crops.

Indeed, the crop biotechnology sector has defined the safety of crops at all points in their development, and notable studies show no difference between genetically modified and non-genetically modified crops apart from the specific gene or genes introduced to the crop.[49] Genetically modified wheat varieties that provide significant opportunities to the bakery industry with regard to protein quality and micronutrition have been shown by research to show no difference to non-genetically modified varieties other than the introduced gene, but their introduction to the food chain is hampered by what can be considered consumer and qualitative issues.[50,51] A notable example of a biotechnology company developing programmes of assessment to communicate the total value of genetically modified or industrially produced organisms is the Novozymes A/S company, which is headquartered in Denmark. Novozymes A/S has used LCA to develop assessments for its products, and these methods assess the benefit of using industrial biotechnology on grounds of sustainability criteria.[52] Such a model of customer engagement and supply chain assessment can provide important insights into how biotechnologies within the food supply chain might be accepted by consumers globally.

1.9   Genetic Diversity of Agricultural Crops and Livestock

Tackling the concerns of introducing new properties and performances to agricultural products raises the whole issue of genetic diversity in the crops and animals we utilise for producing food. Whereas the diversity of product taste and texture has often been introduced by processing and manufacturing new ingredients and foods so that they provide desired performance in recipes, such approaches have not been used in farming. Naturally, the opportunities to do so are limited and were defined by Professor Jack Harlan nearly half a century ago when Borlaug was beginning to formulate the route of the first Green Revolution. The drivers for this first Green Revolution were genetic diversity, and Harlan's reconsideration of the origins of crops and livestock were important in this context. Harlan redeveloped existing ideas on biodiversity and agriculture in that he established that the world food system depends on a limited number of plant species and these evolved in three areas of agriculture that include the Near East, Mesoamerica and North East Asia.[53]

Previous analyses considered between 8 and 12 centres of origin, which Harlan called non-centres because of the variation in genetic diversity within regions from which crops and livestock first developed, were not strictly defined as centres because they were diffuse. Regardless of the actual centres or areas of origin of agriculture, it is clear that the genetic stock of agricultural systems is limited because of the relatively few species utilised and the narrowly defined geographic regions they first developed from. Whereas the number of crops that supply most calories globally number under 10 species (mainly cereals), the number of plant species utilised for taste, fragrance, and medicine is likely to reach several thousand, with some estimates around 50 000 species. These metrics provide an illustration of the complexity in relating crop and livestock diversity to specific centres of origin or species, because it is highly likely that agriculture evolved with complex relationships within landscapes. Thus, while it is useful to think of centres of origin for agriculture, we must consider the genetic diversity that is used for the whole food system in geographic regions, including taste, fragrances, and medicines.

1.10   Trade Agreements and the Development of Agricultural Supply

If we are to consider the efficiency of the agricultural production system, which is the producer of food and animal feed ingredients, then the development of trade agreements globally have without a doubt been most influential in enabling new technologies and management. In Europe, the development of the Common Agricultural Policy (CAP) has dominated the agricultural sector since the late 1950s (Figure 1.7).

The development of the CAP provides an important case study in the development of a sustainable agricultural system. This is because several reforms or adaptive changes

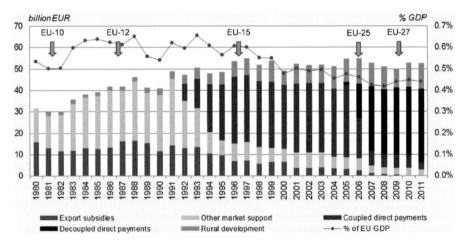

Figure 1.7.   CAP expenditure and reform path budget evolution; while overall (unadjusted) budget has increased, it is worth noting the changes in distribution of the main measures (specifically the increase in Direct Aids and Rural Development), as well as the significant increase in number of farm businesses, together with the enlargement of the EU.
Source: **The graph is from the DG Agriculture and Rural Development, Agricultural Policy Analysis and Perspectives Unit © European Union (2013). http://ec.europa.eu/agriculture/cap-post-2013/graphs/index_en .htm (accessed 22 April 2014). (For a colour version, please see the colour plate section.)**

have been made from initially aiming to develop stable markets by price support to developing sustainable landscapes and a resilient food system. In the 1980s, CAP expenditure was targeted to price support through market mechanisms, such as intervention (when world prices were low) and export subsidies. Market subsidies resulted in agricultural surplus, and in 1992, there was reform of the CAP, where market mechanisms were reduced and replaced by direct payments. Direct payments were 'producer support' not 'price support' for agricultural products. Further reform has increased spending on rural development measures that enhance landscapes and ecosystem services.

Figure 1.7 also shows the impact of the 2003 reform where direct payments were shifted to decoupled payments where CAP payments are no longer paid per hectare or per animal but made in terms of farming function. This is related to sustainability and environmental criteria associated with farming, as well as the functions of biomass production for the food system. Spending on rural development has developed during this period, showing an important consideration for the food production to be integrated into the social and community well-being of society. CAP spending has stabilised despite the successive enlargements and represents 0.44% of the European Union gross domestic product (GDP). However, the CAP expenditure as share of the EU budget has decreased over the past 25 years, from almost 75% to 44% in 2011. This decrease has taken place despite the successive EU enlargements and the downward path of CAP cost in the EU is due mainly to the CAP reforms.

References

1   Martindale, W., & Lillford, P. (2008). Will an innovative and sustainable food system supply nine billion shoppers? *Aspects of Applied Biology, 87*, 43–44. http://www

.foodinnovation.org.uk/download/files/LILLFORD
.pdf (accessed on 30 April 2014).

2 Ingram, J. S. I., Wright, H. L., Foster, L., Aldred, T., Barling, D., Benton, T. G., Berryman, P. M., Bestwick, C. S., Bows-Larkin, A., Brocklehurst, T. F., Buttriss, J., Casey, J., Collins, H., Crossley, D. S., Dolan, C. S., Dowler, E., Edwards, R., Finney, K. J., Fitzpatrick, J. L., Fowler, M., Garrett, D. A., Godfrey, J. E., Godley, A., Griffiths, W., Houlston, E. J., Kaiser, M. J., Kennard, R., Knox, J. W., Kuyk, A., Linter, B. R., Macdiarmid, J. I., Martindale, W., Mathers, J. C., McGonigle, D. F., Mead, A., Millar, S. J., Miller, A., Murray, C., Norton, I. T., Parry, S., Pollicino, M., Quested, T. E., Tassou, S., Terry, L. A., Tiffin, R., van de Graaf, P., Vorley, W., Westby, A., & Sutherland, W. J. (2013). Priority research questions for the UK food system. *Food Security*, 5, 617–636.

3 Meadows, D. H., Meadows, D. L., Randers, J., & Behrens, W. W. (2009). The limits to growth. *The Top 50 Sustainability Books*, 1(116), 31–37.

4 Rosenzweig, C., & Parry, M. L. (1994). Potential impact of climate change on world food supply. *Nature*, 367(6459), 133–138.

5 Piesse, J., & Thirtle, C. (2009). Three bubbles and a panic: an explanatory review of recent food commodity price events. *Food Policy*, 34(2), 119–129.

6 Headey, D., & Fan, S. (2008). Anatomy of a crisis: the causes and consequences of surging food prices. *Agricultural Economics*, 39(s1), 375–391.

7 Godfray, H. C. J., Beddington, J. R., Crute, I. R., Haddad, L., Lawrence, D., Muir, J. F., Pretty, J., Robinson, S., Thomas, S. M., & Toulmin, C. (2010). Food security: the challenge of feeding 9 billion people. *Science*, 327(5967), 812–818.

8 Headey, D. (2011). Rethinking the global food crisis: the role of trade shocks. *Food Policy*, 36(2), 136–146.

9 Costanza, R., d'Arge, R., De Groot, R., Farber, S., Grasso, M., Hannon, B., Limburg, K., Naeem, S., O'Neill, R. V., Paruelo, J., Raskin, R. G., Sutton, P., & Van den Belt, M. (1997). The value of the world's ecosystem services and natural capital. *Nature*, 387(6630), 253–260.

10 World Commission on Environment and Development (1987). *Our common future.* New York: Oxford Paperbacks.

11 UN General Assembly (2000). *United Nations millennium declaration.* United Nations General Assembly Resolution 55/2. New York: United Nations.

12 UN General Assembly (2002). *Road map toward the implementation of the United Nations Millennium Declaration.* United Nations General Assembly Document A56/326. New York: United Nations.

13 Haines, A., & Cassels, A. (2004). Can the millennium development goals be attained? *BMJ (Clinical Research Ed.), 329*(7462), 394.

14 United Nations (2013). Millennium development goals report 2013. Available from the official United Nations sites for MDG indicators. http://www.un.org/millen niumgoals/pdf/report-2013/mdg-report-2013-english .pdf (Accessed 30 April 2014).

15 Shoaf Kozak, R., Lombe, M., & Miller, K. (2012). Global poverty and hunger: an assessment of millennium development goal# 1. *Journal of Poverty, 16*(4), 469–485.

16 Anríquez, G., Daidone, S., & Mane, E. (2013). Rising food prices and undernourishment: a cross-country inquiry. *Food Policy, 38,* 190–202.

17 Ivanic, M., Martin, W., & Zaman, H. (2012). Estimating the short-run poverty impacts of the 2010–11 surge in food prices. *World Development, 40*(11), 2302–2317.

18 Von Braun, J. (2007). *The world food situation: new driving forces and required actions.* Washington, DC: International Food Policy Research Institute.

19 Mitchell, D. (2008). A note on rising food prices. World Bank Policy Research Working Paper Series, Vol. Policy Research Working Paper 4682. World Bank (2008) Open Knowledge Repository. https://openknowledge .worldbank.org/handle/10986/6820 (Accessed 30 April 2014)

20 FAO, WFP, and IFAD (2012). *The State of Food Insecurity in the World 2012. Economic growth is necessary but not sufficient to accelerate reduction of hunger XE 'hunger' and malnutrition.* Rome: FAO.

21   von Grebmer, K., Ringler, C., Rosegrant, M. W., Olofin-biyi, T., Wiesmann, D., Fritschel, H., Badiane, O., Torero, M., Yohannes, Y., Thompson, J., von Oppeln, C., & Rahall, J. (2012). *2012 Global Hunger Index: the challenge of hunger: ensuring sustainable food security under land, water, and energy stresses*. Washington, DC : International Food Policy Research Institute; Dublin: Concern World Wide; Bonn: Welthungerhilfe and Green Scenery.

22   Bateman, I. J., Harwood, A. R., Mace, G. M., Watson, R. T., Abson, D. J., Andrews, B., Binner, A., Crowe, A., Day, B. H., Dugdale, S., Fezzi, C., Foden, J., Hadley, D., Haines-Young, R., Hulme, M., Kontoleon, A., Lovett, A. A., Munday, P., Pascual, U., Paterson, J., Perino, G., Sen, A., Siriwardena, G., van Soest, D., & Termansen, M. (2013). Bringing ecosystem services into economic decision-making: land use in the United Kingdom. *Science, 341*(6141), 45–50.

23   Daniel, T. C., Muhar, A., Arnberger, A., Aznar, O., Boyd, J. W., Chan, K. M., Costanza, R., Elmqvist, T., Flint, C. G., Gobster, P. H., Grêt-Regamey, A., Lave, R., Muhar, S., Penker, M., Ribe, R. G., Schauppenlehner, T., Sikor, T., Soloviy, I., Spierenburg, M., Taczanowska, K., Tam, J., & von der Dunk, A. (2012). Contributions of cultural services to the ecosystem services agenda. *Proceedings of the National Academy of Sciences, 109*(23), 8812–8819.

24   World Bank. Development Data Group (ed.). (2012). *World development indicators 2012*. Washington, DC: World Bank.

25   FAO (2014). FAO World Programme for the Census of Agriculture (WCA). http://www.fao.org/economic/ess/ess-wca/en/ (accessed on 22 April 2014).

26   Brandt, W. (1980). *North-South, a programme for survival: report of the Independent Commission on International Development Issues* (Vol. 1). London: Pan World Affairs; Macmillan.

27   Diamond, J. M. (1998). *Guns, germs, and steel: a short history of everybody for the last 13,000 years*. London: Random House

28   Joint FAO (1985). Energy and protein requirements: report of a joint FAO/WHO/UNU expert consultation;

energy and protein requirements: report of a joint FAO/ WHO/UNU expert consultation (no. 724). World Health Organization. Updated by Joint FAO. (2007). Protein and amino acid requirements in human nutrition: report of a joint FAO/WHO/UNU expert consultation.

29 Krebs, J. R. (2009). The gourmet ape: evolution and human food preferences. *The American Journal of Clinical Nutrition, 90*(3), 707S–711S.

30 Simpson, S. J., & Raubenheimer, D. (2005). Obesity: the protein leverage hypothesis. *Obesity Reviews, 6*(2), 133–142.

31 Weisenfeld, U. (2012). Corporate social responsibility in innovation: insights from two cases of Syngenta's activities in genetically modified organisms. *Creativity and Innovation Management, 21*(2), 199–211.

32 Potrykus, I. (2012). 'Golden rice', a GMO-product for public good, and the consequences of GE-regulation. *Journal of Plant Biochemistry and Biotechnology, 21*(1), 68–75.

33 Rawat, N., Neelam, K., Tiwari, V. K., & Dhaliwal, H. S. (2013). Biofortification of cereals to overcome hidden hunger. *Plant Breeding, 132*(5), 437–445.

34 Ehrlich, P. R., & Ehrlich, A. H. (2009). The population bomb revisited. *The Electronic Journal of Sustainable Development, 1*(3), 63–71.

35 The World Food Prize Foundation (2014). The world food prize. http://www.worldfoodprize.org/ (accessed 17 September 2013).

36 Hillel, D. (1991). *Out of the earth: civilization and the life of the soil.* Berkeley: University of California Press.

37 Schlenker, W., & Roberts, M. J. (2009). Nonlinear temperature effects indicate severe damages to US crop yields under climate change. *Proceedings of the National Academy of Sciences, 106*(37), 15594–15598.

38 Lindenmayer, D. B., & Likens, G. E. (2009). Adaptive monitoring: a new paradigm for long-term research and monitoring. *Trends in Ecology and Evolution (Personal Edition), 24*(9), 482–486.

39 Lindenmayer, D. B., Likens, G. E., Haywood, A., & Miezis, L. (2011). Adaptive monitoring in the real world:

proof of concept. *Trends in Ecology and Evolution (Personal Edition)*, 26(12), 641–646.

40   Evans, L. T. (1998). *Feeding the ten billion: plants and population growth.* Cambridge, UK: Cambridge University Press.

41   Parfitt, J., Barthel, M., & Macnaughton, S. (2010). Food waste within food supply chains: quantification and potential for change to 2050. *Philosophical Transactions of the Royal Society B: Biological Sciences*, 365(1554), 3065–3081.

42   Martindale, W. (2014). Using consumer surveys to determine food sustainability. *British Food Journal*, 116(7), in press.

43   Nestel, P., Bouis, H. E., Meenakshi, J. V., & Pfeiffer, W. (2006). Biofortification of staple food crops. *The Journal of Nutrition*, 136(4), 1064–1067.

44   Potrykus, I. (2001). Golden rice and beyond. *Plant Physiology*, 125(3), 1157–1161.

45   Paine, J. A., Shipton, C. A., Chaggar, S., Howells, R. M., Kennedy, M. J., Vernon, G., Wright, S. Y., Hinchliffe, E., Adams, J. L., Silverstone, A. L., & Drake, R. (2005). Improving the nutritional value of Golden Rice through increased pro-vitamin A content. *Nature Biotechnology*, 23(4), 482–487.

46   Tang, G., Qin, J., Dolnikowski, G. G., Russell, R. M., & Grusak, M. A. (2009). Golden Rice is an effective source of vitamin A. *The American Journal of Clinical Nutrition*, 89(6), 1776–1783.

47   Al-Babili, S., & Beyer, P. (2005). Golden rice: five years on the road—five years to go? *Trends in Plant Science*, 10(12), 565–573.

48   Miflin, B. J. (2000). Crop biotechnology. Where now? *Plant Physiology*, 123(1), 17–28.

49   König, A., Cockburn, A., Crevel, R. W. R., Debruyne, E., Grafstroem, R., Hammerling, U., Kimber, I., Knudsen, I., Kuiper, H. A., Peijnenburg, A. A., Penninks, A. H., Poulsen, M., Schauzu, M., & Wal, J. M. (2004). Assessment of the safety of foods derived from genetically modified (GM) crops. *Food and Chemical Toxicology*, 42(7), 1047–1088.

50 Baker, J. M., Hawkins, N. D., Ward, J. L., Lovegrove, A., Napier, J. A., Shewry, P. R., & Beale, M. H. (2006). A metabolomic study of substantial equivalence of field-grown genetically modified wheat. *Plant Biotechnology Journal*, 4(4), 381–392.

51 Shewry, P. R., Baudo, M., Lovegrove, A., Powers, S., Napier, J. A., Ward, J. L., Baker, J. M., & Beale, M. H. (2007). Are GM and conventionally bred cereals really different? *Trends in Food Science & Technology*, 18(4), 201–209.

52 Nielsen, P. H., Oxenbøll, K. M., & Wenzel, H. (2007). Cradle-to-gate environmental assessment of enzyme products produced industrially in Denmark by Novozymes A/S. *The International Journal of Life Cycle Assessment*, 12(6), 432–438.

53 Harlan, J. R. (1971). Agricultural origins: centers and noncenters. *Science*, 174(4008), 468–474.

2 Understanding Food Supply Chains

2.1 Current Methods of Assessing Food Supply Chain Efficiencies That Enable Food Security Projections

This book aims to develop a framework for developing food security worldwide using data and knowledge held by whole supply chains. The twenty-first century has seen the emergence of big data scenarios in that we can obtain volumes of data regarding supply, consumption, and purchase of foods like never before. The food system itself is now at a point where we need to assess the importance of these data and how we can use them to provide sustainable and secure outcomes for food supply globally. The impetus to do this has been made clear by many writers, who are not the focus of this particular study; this book aims to make the reader aware of these writers and commentators and place them into an integrated context so that the reader can make rational decisions that will develop leadership in food security. This cannot be achieved without considering the role of the policymaker, who is anyone who can plan a vision where security and sustainability of food is defined by projecting the outcomes of supply chain activities so that we can plan for a sustainable future.

A dominant theme within policy development is changing goals, and within the food system, it has become a

---

Global Food Security and Supply, First Edition. Wayne Martindale.
© 2015 John Wiley & Sons, Ltd. Published 2015 by John Wiley & Sons, Ltd.

cacophony of changing messages, different values, and continuing demand for a supply chain that can deliver over consumption as the notion of a sustainable meal. From this arena of what sometimes can only be described as noise, a policymaker who aims to change actions for the better must pick out attributes that can help determine a route for what can be regarded as secure and sustainable consumption food policy. The melting pot that is the food system contains both privately controlled and publically controlled elements that can confound any attempts by enlightened and innovative solutions to develop sustainable practices. This is why the food industry has always struggled with profit, assurance, traceability, and contamination. The supply chain has demand attributes that are often focused on analysis of how population growth and higher incomes affect consumption of foods. These relationships are well defined in economics, but they often only consider the outcome of sating demand in populations and rarely consider why demand increases in the first place. In the context of food security, we must largely consider the role of agriculture, but mining of minerals as nutrients will have a crucial role in the development of populations and demand.

2.2   How Population Growth and Limiting Factors Define Demand and Food Security

No discussion of population growth and demand can go ahead without a consideration of Malthusian dynamics that still dominate much of the thinking in the food system. Thomas Malthus realised in the eighteenth century that exponential population growth was possible when resources were available to allow it. This will ultimately create an overshoot in growth that in turn results in a decline in population growth if technological or social innovations do not respond to cater for larger populations.[1] The second half of the twentieth century saw a resurgence in credibility for the Malthusian world view, and many commentators interpreted increasing population growth to mean an unsustainable future scenario. Professor Paul Ehrlich's essay and later book, *The Population Time Bomb*, expounded a Malthusian view of the future and brought Malthusian thinking into the food security debate.[2] Other notable projections came from the Meadow's projections on the limits to growth at this time. They are notable because they were presented to the Club of Rome, which is associated with development of the current European Commission (EC).[3] The EC itself has become an important source of policy for agriculture, and the impacts of agriculture on society and the Meadows' projections demonstrated that the overshoot of populations with respect to resources may happen, with catastrophic impacts on societies.

The greatest weaknesses of the Malthusian world views are they have never happened globally with the impacts that the advocates of Thomas Malthus projected. Naturally, they would say that this is because they have influenced policy to consider impact as well as growth. There is no doubt that Malthusian principles have contributed to the sustainability debate, but we must question why humankind has always managed to provide some form of regulation between the supply of resources and demand of populations.[4] If we are to consider the limits of growth and place them into the context of population projections, it is not simply a case of stating Malthus got it wrong; many commentators have shown this is too simplistic and simply not the case.[5] It is certainly the case that all Malthusian projections tend to overlook the power of adaptive modelling and the introduction of new technologies that will overcome limits by identifying new materials and technologies that will result in more efficient use of current materials. This has evidently been the case for the global food system, where humanity can now control genetic performance and production of biomass.[1] Professor Tony Trewavas has developed a viewpoint based on current capacities of agricultural technologies that seriously question the limits that Ehrlich and others identify.[6] His understanding of new technologies is well grounded in a career that has used biotechnology to decipher how calcium is used by plants to sense environmental stress, but it is the foiling of Malthusian principles that concerns us here. Trewavas points to the delay and misunderstanding of biotechnology as providing limits to current agricultural technologies. His approach is proving correct, as many governments are beginning to reevaluate the role of biotechnology in agriculture. However, he also acknowledges the need to understand the qualitative nature of the biotechnology solutions and their interaction with consumers, warning that the 'cult of the amateur' will result in an unjust representation of the biotechnology industry, which has become completely transparent to the consumer.[7]

2.3 Global Population Estimates and Projections

The total global population estimate has consistently increased since industrialisation. These figures represent a global average and there are drastic differences between nations and regions of the world with regard to the amount of population growth. Analysis of population growth generally shows increased rates of growth in developing regions of the world and stable growth in developed regions. The average global trend also hides a significant move from rural population growth to urban population growth, with the later increasing significantly over rural population growth. This is an important trend because population becomes more concentrated in urban areas, requiring intensification of food production and supply mechanisms. The impact of population growth and distribution is a key consideration in environmental quality impacts. When populations become more concentrated and their requirement for efficient food supply more apparent, the need for assessing the impact of agricultural operations will become more important. There are a number of methods for doing this, including the development of indicator frameworks.

Supplying and increasing world population with due consideration of the capacity of the food system is an important international concern, and the provision of a productive agricultural systems to do this is essential. This must be achieved within the bounds of political, economic, and technical abilities that are always changing. For example, a technical attribute, such as crop nutrition, is an important indicator of agri-system efficiency and the supply of food to the global population. Crop nutrition is a component of producing food that can be represented in terms of crop yield, calorie sufficiency, and quantity of foodstuffs. It is also an important component of food quality, with trace element nutrition being vital to the production of food with high nutritive quality as well as calorific content.

The complexity of the food sustainability goal is embodied in the impact of continued global population growth and cultural transitions. The United Nations population projections and the UN Food and Agricultural Organisation food production and consumption trends provide a means to develop an assessment regarding the question of how much food needs to be produced in the next 40 years. Research presented by Keating and Carberry (2010) provide scenarios based on low and medium world population projections of 8.0–9.0 billion in 2050. They have assumed human fertility trends and consumption remain constant based on previous population data. The scenarios show the world will need to produce over 380 to over 400 exa calories (1 exa calorie $= 10^{18}$ calories) over the period of 2000 to 2050 if an average consumption of

2255 kcal/person/day is maintained. This is equivalent to the food the world has produced over the 200 years pre-2000 in order to feed its population.

A further scenario presented by Keating and Carberry (2010)[8] assumed that a mean global consumption of 3590 kcal/person/day by 2050 is reached. This might be considered more realistic if current consumption trends continue and the resulting demand will be for over 450 exa-calories, which is the equivalent to the food produced in 330 years pre-2000. The scenarios presented provide an assessment of the challenge that faces food supply over the next 30 years; in reality, we will have to produce food at an efficiency which is up to 10-fold greater than we are currently doing. This means that the agricultural segment of the food supply chain is critical to producing these calories required by the growing demands of populations. Population growth is tracked and projected using models that have become increasingly robust, but consumption of food is also dependent on the cultural values, changes in individual taste preference, and trends in food category popularity. When these cultural variables are interfaced with supply–demand functions, the food system is clearly extremely complex and not just about supplying calories and protein.

2.4   Consumption and Population Growth: Demonstrating the Impact of Dietary Changes and Transitions

The food security implications of population projections become even more profound when we relate them to the consumption of protein. This is because 65% of global protein consumption is from just seven major food ingredients, as reported by the FAO statistical database, FAOSTAT. These are wheat (20%), rice (12%), maize (5%), dairy (10%), beef (6%), poultry (6%), and pork (6%).

Furthermore, livestock product consumption provides specific stressors on the global food system, because feed protein is consumed by livestock to produce meat as a protein source at conversion efficiencies that range from 5% for beef products to 40% for dairy products.[9] The nutrient transition of the world food system to more meat-containing diets has resulted in an increased demand for meat, creating a dilemma for sustainability of resource use because of the demand for feed protein that directly diverts protein from food supply chains.[10,11]

This can be demonstrated using current world production of meat reported by FAOSTAT in 2011, which is nearly 300 million tonnes. If we assume that a typical tonne of meat will need at least 10 tonnes of feed to produce it and this feed is wheat that achieves a world average yield of 3 tonnes for each hectare grown then we require 3 billion tonnes of wheat or 1 billion hectares to produce it. While I accept these are very much typical calculations, they do present important principles. If the reader needs to see detailed analysis regarding capacities of the food production system that exist and are utilised by policymakers, then investigating other projections that have been published will also show the principle and value of identifying baselines we can project scenarios from.[12] We know that feed conversion into meat ranges from 2.3 kg (for fish), 4.2 (for chicken), 10.7 (for pork), and 31.7 (for beef) because Smil has reported these figures previously using USDA data.[9] While these figures can be improved, they do represent an important reference point without carrying out detailed life cycle assessment (LCA) of crop to feed conversion in livestock systems. An energy balance approach is now discussed here as an important approach because it enables us to appreciate the scale of the meat supply issues. We would need 1 billion hectares of cereal production to feed the livestock currently produced globally if cereals were the main source of feed. The actual area of wheat harvested in 2012 was 217 million hectares, and of course the feed to animal production system is clearly not that straight-forward. Animal feeds include all cereals, where FAOSTAT reports 703 million hectares were harvested in 2012, and for oil crops, a further 281 million hectares were harvested.

Obtaining the near one billion hectares for livestock feed seems to have a complete solution here and the land resource limitation is overcome, but this is just for animal feed and does not include food, so we still have a problem. Of course this problem is solved by using grassland as grazing systems and forage production, that is, without the consideration of managing grazing and forages, the world livestock system would not work. They are essential to the world food system, and in 2011, FAOSTAT reports that there are nearly 3.4 billion hectares of grasslands globally, which are the engine for livestock production. These are the critical production considerations for the feed that supports a global system of producing grazing animals and livestock products. If we consider FAOSTAT data, there are 0.5 billion hectares of arable land globally; the demand for livestock feed will create increasing pressures on feed protein supply and the food system. Previous interventions to supplement feed protein have focused on improving grazing systems so that the 3 billion hectares of permanent pasture available globally can efficiently achieve optimal grazing and forage production to reduce demand for cereal and oilseed feeds. However, there is the increasing importance of processors and manufacturers in providing a means to ameliorate pressure on the meat consumption system by the utilisation of all co-product and waste streams for protein supply. Furthermore, the growing importance of using industrially produced protein that converts starch to protein at much higher efficiencies than livestock systems is emerging in response to these pressures.

The role of processors and manufacturers to convert biomass into different protein product ranges are critical to future conservation of livestock resources, and it can provide a basis for defining sustainable meal and diet planning for populations. The delivery of sufficient protein for human well-being is a human right, and the food processing and manufacturing industry offers many opportunities for providing routes to efficient feed protein and livestock protein production, and industrial starch to protein conversions. The processing industry is of huge importance in optimising protein consumption through the design of recipes, meals, and products so that consumer well-being is enhanced and environmental impacts are reduced. Relating product development to whole meals and diets is critical, and it requires a full appraisal of whole supply chain activities. As previously highlighted, processors and manufacturers operate in the middle of the supply chain and as such have an important place in determining traceability of food ingredients to assure safety or provide robust responses in crisis management. Naturally, one of the most important considerations for food supply chain management is population growth, which is further compounded by increasing urbanisation and nutrient transition trends. With this in our minds, we must start to determine what food supply functions result in security, and the nutrients used in agriculture are an important place to start because they ultimately determine the nutritive quality of the biomass produced by agricultural systems. This biomass can be enhanced by processing and manufacturing, but ultimately the quality of biomass produced will be an important determinant of eventual nutritive value of foods.

2.5 Optimising Nutrition across Supply Chains Is the Focus of the Second Green Revolution

As we have seen how the work of Borlaug, Evans, and Harlan have transformed how new crops are designed with respect to production demands, we must consider not only producing more food, but also more nutritious foods that are tailored for specific nutritional requirement. The world demand for specialised dietary products is likely to increase as global diets become calorie sufficient or, more worrying calorie over-sufficient. How this food quality goal in diet is distributed globally will be an important issue in the twenty-first century. The distribution of food in the global system has been characterised and we now have traceability standards associated with food categories that assure consumers.[13] Whereas food distribution is characterised because of the requirements of food safety and traceability issues associated with supply chains, there is increasingly a need to qualify the sustainability of distribution with respect to environmental impacts and loss of nutritive value.[14,15]

Some commentators have said that this is the basis of a second 'green' revolution, where the first aimed for calorific- and supply-based goals, and the second will aim for these actions alongside qualitative goals associated with nutritional values of foods. These will be more closely linked to improving health of nations that have reached calorific sustainability. Nutrient availability and consumption have a huge effect on the ability to produce food of enough quantity and quality to support industrialised societies. The manner in which culture and societies develop in specific areas may not always be associated with the most favourable areas for food production, and the import of nutrient sources into these areas using efficient distribution networks becomes very important. The development of efficient supply chains has extreme social and economic impacts on national development programs and can create very specific environmental impacts.

The need for nutrients to be economically assessed by producers at the beginning of the food supply chain is of extreme importance; they must consider the yield potential of a particular landscape and soil with the benefits of using optimal amounts of nutrients for increased biomass yields. This must then be equated against the cost of nutrients and their distribution and application for achieving optimal yield. Integrated Plant Nutrition Systems (IPNS) emphasise the importance of economic assessments being made that align with agronomic yield potential, so that the economic and technological factors for production are taken into account together, not as separate factors. Environmental assessment of farm inputs such as plant nutrients is an important consideration, and it can be formalised as a farm nutrient management plan because the ecology for a particular landscape can be significantly changed by different farming practices. For example, reduction of wind- and water-enhanced soil erosion has clear economic benefits in conserving the nutrients in top soils. Soil conservation and landscape management are key components of IPNS, and by enacting soil conservation procedures, farmers can ensure environmental and regulatory compliance.

2.6 The Emergence of Sustainable Farming Reconnecting Supply Chains: A Case Study of the Establishment of the Landcare Movement in Australia

These types of approaches to landscape management described here were perhaps first placed into farming systems and the food supply chains associated with them by the founders of the Australian Landcare movement at the end of the 1980s in South West Victoria. Sue Marriott and Andrew Campbell, among others, started this by providing a forum for groups of farmers to discuss farm planning that improved business operations and developed sustainability criteria associated with farms, such as water and soil conservation.[16] The Landcare movement started with a number of farms working together to share management information in a project supported by the Ian Potter Foundation; they became known as the 'Potter Farms'. Management systems, such as the sward management system, Prograze™, developed from these actions, and John Marriott pushed forward a programme of farm benchmarking to improve business performance with respect to economic and sustainability criteria.[17] The initial decade of the Landcare movement demonstrated that important tasks could be met by farmers worldwide in obtaining a sustainable and secure food system.[18,19] At its heart was an understanding of natural resources and how the use of soil management, plant nutrients, crop and livestock protection, and water and energy balance could be used to develop a farming system to be proud of.[20] The future challenges were to integrate this with the food supply chain and land use industries, which to some extent defines where the second green revolution goals are, that is, looking towards the consumer components of the supply chain.

2.7 The Long-Term Field Experiments at Rothamsted and Their Power of Demonstrating Good Nutrient Balance in Agriculture Has Been Crucial to the Development of Sustainable Food Supply

In order to grow and develop satisfactorily, all plants need a supply of carbon, hydrogen, and oxygen, which they get from soil, air, and water environments. There are also 13 essential mineral elements (nutrients) shown in Table 3.1. These elements are normally obtained by plants from the soil, and it is useful to reflect on where our current understanding of soil fertility derives from. The facts that provide these foundations for modern agronomy were initially researched by Justus von Liebig in the early to mid-nineteenth century and applied by Sir John Bennet Lawes of Rothamsted and Sir Joseph Henry Gilbert of Rothamsted in initiating the modern fertiliser industries. Liebig made measurements of how dry matter was gained in plant systems and hypothesised that much of the nutrient required for growth would come from the soil but he thought nitrogen was essentially obtained from the atmosphere. These ideas were later developed by John Lawes, the founder of the Rothamsted Long-Term Agricultural Experiments, and Henry Gilbert, the chemist who worked with Lawes to develop the field trial database that still exists today.[21] Their experiments have also established our current understanding of the nitrogen cycle, which shows plants obtain nitrogen from soils and its bioavailability is critical to the production of all biomass.[22]

All natural resources are finite and optimising plant nutrient supplies for an ever-growing global population remains as important today as ever, and this provides a critical role for the data obtained from long-term experiments, such as those at Rothamsted.[23] The maintenance of the amount of food produced and the nutritional value of food will be dependent of the minerals supplied to crops during production. This means that the importance of plant nutrition is one of the most important components of production. The consumption of nitrogen (N), phosphate (P), and potassium (K) globally reflects the importance of providing enough nutrients to support an efficient food production system.[24,25] The historical evidence obtained from the Rothamsted trials can demonstrate the sustainability of crop production in the arable farming systems that produce cereals, maize, grain legumes, and rice, which are the dietary powerhouses crucial to delivering the calories and protein for human diets (see Figure 1.4). The sustainable or continued production of these crops is extremely important to global agricultural production, and IPNS will be a key component of achieving agri-sustainability in these systems.

The use of long-term data from the global long-term experiments such as those of the Morrow Plots in the

United States and those at Rothamsted in England have provided important lessons for the sustainability of grain-based production systems by demonstrating how it can be achieved. These experiments are over 100 years old, and they have shown that using the optimal rate of nutrient inputs generally corresponds to obtaining the optimal yield of a crop. This, combined with the use of crop rotations and break crops, can improve nutrient use efficiency, soil organic matter content (SOM), and reduce disease pressure as compared with continuous cropping.[26] The impact of weed pressures has also been investigated in the long-term agricultural experiments and provides important data that describe the emergence of herbicide resistance in weed species, which is now having serious impacts on cereal yields.[27] The interaction between crop yield limitations and crop varieties has also been explored by benchmarking modern varieties to older varieties so that yield benefits observed as a result of introducing new varieties and crop breeding can be identified to be between 10% and 30% of current yields.[28]

2.8   Long-Term Field Experiments Hold Critical Data That Provide Our Understanding of Nutrient Flows in Farming Systems So That Sustainable Food Supply Chains Are Developed

The global long-term experiments have provided valuable data on the soil organic matter (SOM) content of agricultural soils, which are critical to understanding global carbon and nutrient cycles. It has been found that cultivation of the soil will decrease organic matter content of most soils, and this decrease will reach an equilibrium level that is associated with the particular soil and cultivation system being investigated.[29] This equilibrium level will largely depend upon agronomic, climatic, and soil structural factors. Conservation of SOM can be achieved by optimal soil cultivation methods (such as minimal tillage or no till where possible), managing animal manure, and organic matter inputs utilising crop rotations effectively. The correlation between optimal soil organic matter, soil type, and long-term fertility is a strong one but very difficult to define in many field situations. Farm-recorded data collected over many rotations and several seasons are important in assessing the optimal organic matter content for a given soil because it is a means to maintain soil fertility and conserve the GHG emissions associated with farming. This is one of the reasons record keeping over long-term time series is essential to determining what sustainable production is.

How major plant nutrients interact with the soil and SOM is also critical to their bioavailability and accumulation by crops. Nitrogen is a major nutrient that is usually added at amounts of 100–250 kg/ha/yr to supplement soil supplies that are made available by mineralisation.[30] Mineralisation is the result of microbial metabolism whereby organic forms of nitrogen that include proteins and amine compounds in the soil are converted to inorganic forms, such as ammonium and nitrate ions, which are soluble, and nitrate ions are both soluble and mobile in soils.[31,32]

Phosphate and potassium are the other major nutrients most important to establishment of crops in soils. Large amounts of these nutrients can be available over varying timescales because they are less mobile and soluble in water than nitrate ions. In general, there are 'nutrient pools' in the soil for P and K that are immediately available, and those that are available over several growing seasons, decades, and even hundreds of years. Thus, the soil fertility with respect to phosphate and potassium can be either accumulative over many years or just maintained by addition of fertilisers. This type of 'nutrient banking' in soils has been shown at the Rothamsted Hoos Field long-term experiment in the United Kingdom, where phosphate additions over 100 years ago still present yield benefits because of the slow release of phosphate from fertilisers added over many years.[33]

Changes in soil pH can radically change the mobility of nutrients in soils, and most soils will decrease in pH over time unless they have a limestone bedrock that is disturbed or alkaline water flows through them.[34] Knowing how to manage these pH changes and pools of nutrients with respect to timescale represents an important management strategy for productive and sustainable farming. Components of the physical breakdown of soils by erosion and weathering will be important in releasing P and K from less available pools but these releases are generally too low to support productive crop production. However, where land has been extensively manured with mineral or organic fertiliser, the long-term value of phosphorus and potassium can be considerable. Many trials have definitively shown that these nutrients can be released to crops over many decades with the Hoos Field at Rothamsted being of highest profile.

Natural events such as the deposition of sulfur and nitrogen nutrients from the atmosphere or from flood waters is a slow process with around 45 kg/ha nitrogen being available and less than 1 kg/ha of phosphorus and potassium being deposited in any year. Flooding can enrich soils significantly with phosphate if there are regular flooding periods, and this will result in considerable spatial variation in nutrients across landscapes and fields. Manure and slurry either from grazing animals or from spreading of stored material is the most important source of plant nutrients from non-industrially produced mineral fertilisers. The purpose of fertilisers and manures are to supplement the natural supply of nutrients required for optimal yield by using fertiliser recommendations. Fertiliser recommendations supplement the nutrition requirement of crops and can be described using principles of nutrient balance, that is, what is removed from land is maintained by fertilisers to meet yield demands.

2.9 The Sustainable Production of Livestock and Long-Term Data

The nutrient balance of sustainable livestock systems deserve special attention, and there are also long-term experiments that consider grassland systems, which include the Park Grass experiment at Rothamsted in England. Park Grass demonstrates interactions between species on long-term grass sward development with different lime and nitrogen, phosphorus, and potassium inputs.[35] This trial has been permanent grass since 1856 and provides a case study for developing species diversity and understanding responses of natural grasslands to lime and nutrient management. An important impact of the experiment has been to show that decreased soil pH on natural grassland is a natural process that can be increased by manure and nitrogen applications.

Liming practices are therefore extremely important in order to produce optimal biomass. In manipulating pH and nutrient status, the species composition of swards can change, and this is an important consideration for biomass production and development of biodiversity. Experiments such as Park Grass show us that the effects of soil pH at nitrogen, phosphorus, and potassium applications must be looked at as a whole to assess the value of a particular management regime. This is a particularly important consideration when considering indicators of biodiversity and environmental quality.

Livestock production system long-term data are somewhat harder to find but trials do exist, such as those of the Palace Leas trials at Newcastle University in England.[36] This trial demonstrates the importance of integrating nutrient inputs from fertiliser, manure, and livestock into nutrient programs, as well as the type of grazing or grass conservation management used. Management, storage, and distribution of organic manure will be an important consideration in any livestock system. Innovative methods of utilising manures are becoming available, including the production of dry manures (particularly from poultry production).[37] The development of efficient manure handling can result in more efficient nutrient use, and the production of forages and conserving grass biomass for feed can result in less nutrients being imported into the grazing system.

Stocking density for particular enterprises will be an essential consideration to prevent overgrazing and erosion of top soil. The management of nutrient inputs must be assessed with the production goals of the farming enterprise in mind, and this is not necessarily the optimal yield. Production and the ensiling or storage of forages produced on farm also reduces the need to import feeds and nutrients onto the farm. The aim of the IPNS system for livestock is to reduce the reliance on nutrient imported into an enterprise. The use of these nutrient budgets can be drawn for livestock enterprises and can be an essential management tool that can be integrated into a nutrient management plan for a given enterprise.[38]

2.10   The Historical Proof of the Value of Agricultural Innovations in Providing Food Security

The history of agronomic breakthroughs shows us that for innovations to be successful, there is often a convergence of investment and technology that results in market entry. Nowhere was this more apparent than the development of the plant nutrient and fertiliser industries in nineteenth-century England.[39,40] The specifics of what occurred are important in the context of deploying any new technology in agriculture since then. The mid-nineteenth century was a time of rapid industrialisation and urbanisation. The problems faced by a growing population were reflected in Sir John Bennett Lawes and Sir Joseph Henry Gilbert's aim of understanding crop production so as to optimise the use of plant nutrients and agronomic techniques to provide food security. They realised that providing increasing population numbers with a sufficient food supply would require an efficient agricultural industry and this increased efficiency could only be reliably demonstrated by agricultural trials.[41]

John Bennet Lawes is the founder of the long-term experiments and the Rothamsted Experimental Station. He was the owner of the Rothamsted estate that he inherited from his parents. Lawes was also one of the first people to manufacture and patent superphosphate, initially using bones that were reacted with sulfuric acid. Lawes developed the superphosphate production into an industry, which generated a substantial fortune for his family. Although he made his fortune producing mineral fertiliser, he advocated the value of recycling nutrients by rotations and the efficient use of organic manure in much of his work. Joseph Henry Gilbert was an experimenter and chemist who provided the technical excellence for the development of Lawes' programs of agronomic research at Rothamsted. Gilbert was driven to obtain indisputable proof and experimental data on which the current understanding of biomass production in the agricultural industry is based.

Gilbert and Lawes worked in partnership for nearly 60 years, laying the foundations for the scientific understanding of crop rotation, soil fertility, and the sustainable use of plant nutrients. By establishing the long-term agricultural experiments at Rothamsted in the United Kingdom during the 1850s, they provide an important step in understanding what sustainable agriculture is. The scientific- and trial-based approach to understanding the soil, water, and atmosphere relationships also led to the development of superphosphate industry in England during the mid-nineteenth century. Phosphate was known to be a limiting factor for crop production on many soils in England, and prior to the production of phosphate fertilisers, the only source was organic manures. In the 1830s and 1840s, John Lawes and others experimented with reacting bone materials with sulfuric acid to make the bone phosphate more available to crops. This resulted in the production of superphosphate, for which Lawes had a patent made in 1843. At this time, around 40 000 tonnes of bones were being imported into England, with a further 26 000 tonnes being produced in England each year for industrial uses. Lawes successfully developed the first fertiliser product with superphosphate, a mixture of calcium phosphate and sulfate, and developed factories by the River Thames in London for their production.[42] Bones were soon replaced with rock phosphates in the manufacturing process, and this had led to overmining guano rich deposits derived from bird droppings and the current debate around the criticality of 'peak phosphate' reserves.

Where Lawes and Gilbert were particularly innovative in their approach was to relate fertiliser products to field trial evidence so that the benefits of using fertilisers responsibly could be clearly seen by farmers. The trials set up at Rothamsted first showed clear crop yield benefits, but in time, further aspects of the plant, soil, and atmosphere system were understood. For example, the way in which plant nutrients became mobile in soils was determined by Lawes and Gilbert when they constructed field drain experiments to determine water balance and nutrient balance of crops. These experiments determined that nitrate remained soluble in soils and it was therefore mobile and leached from soils. This laid down the first understanding of potential nutrient pollution, and at the end of the twentieth century, nitrogen applications to crops are regulated in European Union member countries to reduce nitrate enrichment of water supplies.[43] The field drain experiments at Rothamsted also determined how phosphate and potassium interacted with the water and soil system, providing the first understanding of readily available and slowly available nutrient pools in the soil due to the action of cation exchange.[44] Thus, the Rothamsted trials that Lawes and Gilbert started were not only about demonstrating to farmers the worth of using Lawes fertilisers produced in his factories in East London, but they also established the foundation of understanding nutrient sustainability and innovation in farming systems.

The Rothamsted Classic experiments enabled the development of the fertiliser industry and now they offer important evidence for sustainability landmarks that can be demonstrated by field trials that have collected data since the 1850s. In particular, the development of phosphate fertilisers that not only made the fortunes of the founders but enabled the development of urban populations and reduced the reliance on guano fertilisers mined from South Atlantic islands. Phosphate remains a limiting nutrient globally, and the availability of phosphate reserves is likely to be of increasing importance to the world food production system.[45] Field trial-derived innovations do not obtain market entry easily, as can be seen in Figure 2.1; the fertiliser recommendations identified by Lawes and Gilbert were not able to be fully implemented until the late twentieth century. The reasons for this were one of nutrient availability, regulatory considerations, and the ability to manage new fertiliser materials. The current market entry of genetically modified crops such as golden rice have taken over a decade to reach market so far, and it is perhaps not surprising it may take longer when we consider time lags between invention, product development, and market entry. All of the new technologies in the agricultural system, including nutrient use, liming, and pest control have time lags to market entry. Indeed, the questioning of whether delay is appropriate in a world where we need to increase production for security raises many issues.

The long-term data sets developed by these trials can support methods such as total factor productivity (TFP),

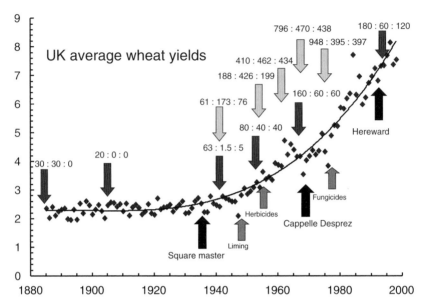

Figure 2.1. The development of agronomic innovations and their relationship to average wheat yields in the United Kingdom. Red arrows show the recommended N:P:K fertiliser application (the fertiliser recommendation for wheat in kilograms per hectare) at that date, the actual amounts of fertiliser as N:P:K applied is shown with blue arrows (in thousand tonnes). Development of different wheat varieties are shown with black arrows, and new agronomic techniques are shown with green arrows. The graph demonstrates the agronomic development of the wheat crops and the innovations that have made dramatic grain yield increases possible.
Source: **The wheat yield data were obtained from FAO. FAOSTAT (2012). Production, crops (data set). http://faostat.fao.org/ (accessed 22 April 2014). Fertiliser data were adapted from the data presented by Cooke (1977),[46] and agronomic practice data were developed by W. Martindale as part of an OECD Cooperative Research Fellowship (2001).[47] (For a colour version, please see the colour plate section.)**

which is used to assess the value of specific parts of the food supply chain, particularly with respect to economic output.[48] These types of methodology can be integrated with measurements of ecosystem services so that they can be applied to the food system globally.[49–51] The data from long-term experiments such as those at Rothamsted in the United Kingdom are crucial to the development of such assessments because historical data improve the projection and understanding of current growing season data. Figure 2.1 demonstrates this because wheat yield increases in the United Kingdom can be related to breakthroughs in plant nutrition, plant nutrient availability, and agronomic innovations.

2.11 The Relationship between Field Trials, Investments, and Innovation

The transfer of field trial research into agriculture and the food supply chain is crucial and can be regarded as the innovation process; that is, the transferring research to market-ready application. For this to happen, field trial research must be placed on farms in a user friendly form that can provide an assurance that optimal decisions are being made with regard to crop management and the food supply chain. As an example of research knowledge to market, the amount of protein in the leaf tissue of crops has been shown to relate to both nitrogen supply and the gas exchange of water vapour and carbon dioxide into and out of leaf tissue.[52] Specific physiological responses, such as the regulation of water balance within a crop, relate to nitrogen supply and protein content of leaves, as well as how farmers will deploy management decisions such as fertiliser application and irrigation. It is therefore not surprising that attempts are made at using the amount of membrane-bound proteins or soluble proteins as indicators of productivity, disease, and stress in crops.[53]

Many environmental stresses experienced by crops will influence the amount and activity of these photosynthetic proteins present in leaves that contain pigments such as chlorophyll. The uses of chlorophyll sensors that have become commercially available are becoming routine to assess crop health and nitrogen fertiliser requirements of crops. The amount of chlorophyll in some plants is closely related to the amount of protein present, and thus chlorophyll can be used as an indicator of nutrient or protein status. Relating this to physiological problems with crops such as water stress and disease is now possible, and the use of metabolic markers such as the 'greenness' of leaves offers much hope in crop diagnostics and managing crop responses to environmental stresses. The importance of now-available chlorophyll meters is that they standardise measurements made in the field so that actions can be guided by robust research and reproducible measurements.[54] These indicators of crop health, such as 'green index', do provide important diagnostic tools for crop management, and it is increasingly important that agronomists have some understanding of crop biochemistry and physiology in order to be able to apply these methodologies.

References

1   Trewavas, A. (2002). Malthus foiled again and again. *Nature, 418*(6898), 668–670.
2   Ehrlich, P. R. (1971). *The population bomb* (Vol. 68). New York: Ballantine Books.
3   Meadows, D. H., Meadows, D. L., Randers, J., & Behrens, W. W. (1972). *The limits to growth: report for the Club of Rome's project on the predicament of mankind*. New York: Earth Island, Universe Book (1974).
4   Daily, G. C., & Ehrlich, P. R. (1992). Population, sustainability, and Earth's carrying capacity. *Bioscience, 42*(10), 761–771.
5   Ehrlich, P. R., & Holdren, J. P. (1971). Impact of population growth. *Science, 171*(3977), 1212–1217.
6   Trewavas, A. J. (2001). The population/biodiversity paradox. Agricultural efficiency to save wilderness. *Plant Physiology, 125*(1), 174–179.
7   Trewavas, A. (2008). The cult of the amateur in agriculture threatens food security. *Trends in Biotechnology, 26*(9), 475–478.

8 Keating, B. A., Carberry, P. S., & Martindale, W. (2010). Sustainable production, food security and supply chain implications. *Aspects of Applied Biology, 102, 7–19.*

9 Smil, V. (2002). Nitrogen and food production: proteins for human diets. *AMBIO, 31*(2), 126–131.

10 Gibbs, H. K., Johnston, M., Foley, J. A., Holloway, T., Monfreda, C., Ramankutty, N., & Zaks, D. (2008). Carbon payback times for crop-based biofuel expansion in the tropics: the effects of changing yield and technology. *Environmental Research Letters, 3*(3), 034001.

11 Zaks, D. P. M., Barford, C. C., Ramankutty, N., & Foley, J. A. (2009). Producer and consumer responsibility for greenhouse gas emissions from agricultural production: a perspective from the Brazilian Amazon. *Environmental Research Letters, 4*(4), 044010.

12 Foley, J. A., Ramankutty, N., Brauman, K. A., Cassidy, E. S., Gerber, J. S., Johnston, M., Mueller, N. D., O'Connell, C., Ray, D. K., West, P. C., Balzer,C., Bennett, E. M, Carpenter, S. R., Hill, J., Monfreda, C., Polasky, S., Rockström, J., Sheehan, J., Siebert, S., Tilman, D. & Zaks, D. P. (2011). Solutions for a cultivated planet. *Nature, 478*(7369), 337–342.

13 Ercsey-Ravasz, M., Toroczkai, Z., Lakner, Z., & Baranyi, J. (2012). Complexity of the international agro-food trade network and its impact on food safety. *PLoS ONE, 7*(5), e37810.

14 Eom, Y. S. (1994). Pesticide residue risk and food safety valuation: a random utility approach. *American Journal of Agricultural Economics, 76*(4), 760–771.

15 Akkerman, R., Farahani, P., & Grunow, M. (2010). Quality, safety and sustainability in food distribution: a review of quantitative operations management approaches and challenges. *Or Spectrum, 32*(4), 863–904.

16 Marriott, S. (2010). Working with Landcare to achieve a sustainable food supply. In W. Martindale (ed.), *Delivering food security with supply chain led innovations: understanding supply chains, providing food security, delivering choice,* Royal Holloway, Egham, UK, 7–9 September 2010,

Aspects of Applied Biology (No. 102, pp. 43–49). Warwick: Association of Applied Biologists.

17   Martindale, W., & Marriott, S. (2004). Integrating education, extension and research for the development of sustainable grazing systems: Australian Landcare and the PROGRAZE™ training programmes. *Bioscience Education* (3). http://journals.heacademy.ac.uk/doi/full/10.3108/beej.2004.03000011 (accessed on 22 April 2014).

18   Youl, R., Marriott, S., & Nabben, T. (2006). *Landcare in Australia*. Melbourne: SILC and Rob Youl Consulting Pty Ltd. http://www.landcaretas.org.au/wp-content/uploads/2012/11/landcare_in_australiaJune08.pdf (accessed on 22 April 2014).

19   Campbell, A. (1994). *Community first: landcare in Australia*. London: Sustainable Agriculture Programme of the International Institute for Environment and Development. http://pubs.iied.org/6056IIED.html (accessed on 22 April 2014).

20   Campbell, A., & Siepen, G. (1994). *Landcare: communities shaping the land and the future*. St Leonards, NSW, Australia: Allen & Unwin.

21   Rasmussen, P. E., Goulding, K. W., Brown, J. R., Grace, P. R., Janzen, H. H., & Körschens, M. (1998). Long-term agroecosystem experiments: assessing agricultural sustainability and global change. *Science*, 282(5390), 893–896.

22   Smil, V. (1997). Global population and the nitrogen cycle. *Scientific American*, 277(1), 76–81.

23   Powlson, D. S., MacDonald, A. J., & Poulton, P. R. (2014). The continuing value of long-term field experiments: insights for achieving food security and environmental integrity. In D. Dent (ed.), *Soil as world heritage* (pp. 131–157). Dordrecht: Springer.

24   Smil, V. (1999). Nitrogen in crop production: an account of global flows. *Global Biogeochemical Cycles*, 13(2), 647–662.

25   Smil, V. (2000). Phosphorus in the environment: natural flows and human interferences. *Annual Review of Energy and the Environment*, 25(1), 53–88.

26 Jeger, M. J., & Pautasso, M. (2008). Plant disease and global change: the importance of long-term data sets. *The New Phytologist, 177*(1), 8–11.

27 Moss, S. R., Storkey, J., Cussans, J. W., Perryman, S. A., & Hewitt, M. V. (2004). Symposium The Broadbalk long-term experiment at Rothamsted: what has it told us about weeds? *Weed Science, 52*(5), 864–873.

28 Austin, R. B., Ford, M. A., Morgan, C. L., & Yeoman, D. (1993). Old and modern wheat cultivars compared on the Broadbalk wheat experiment [Rothamsted Experimental Station (UK)]. *European Journal of Agronomy, 2*(2), 141–147.

29 Aref, S., & Wander, M. M. (1997). Long-term trends of corn yield and soil organic matter in different crop sequences and soil fertility treatments on the Morrow Plots. *Advances in Agronomy, 62*, 153–197.

30 Macdonald, A. J., Powlson, D. S., Poulton, P. R., & Jenkinson, D. S. (1989). Unused fertiliser nitrogen in arable soils: its contribution to nitrate leaching. *Journal of the Science of Food and Agriculture, 46*(4), 407–419.

31 Goulding, K. W. T., Poulton, P. R., Webster, C. P., & Howe, M. T. (2000). Nitrate leaching from the Broadbalk Wheat Experiment, Rothamsted, UK, as influenced by fertilizer and manure inputs and the weather. *Soil Use and Management, 16*(4), 244–250.

32 Glendining, M. J., Powlson, D. S., Poulton, P. R., Bradbury, N. J., Palazzo, D., & Ll, X. (1996). The effects of long-term applications of inorganic nitrogen fertilizer on soil nitrogen in the Broadbalk Wheat Experiment. *The Journal of Agricultural Science, 127*(03), 347–363.

33 Heckrath, G., Brookes, P. C., Poulton, P. R., & Goulding, K. W. T. (1995). Phosphorus leaching from soils containing different phosphorus concentrations in the Broadbalk experiment. *Journal of Environmental Quality, 24*(5), 904–910.

34 Blake, L., Goulding, K. W. T., Mott, C. J. B., & Johnston, A. E. (1999). Changes in soil chemistry accompanying acidification over more than 100 years under woodland and grass at Rothamsted Experimental Station, UK. *European Journal of Soil Science, 50*(3), 401–412.

35 Jenkinson, D. S., Potts, J. M., Perry, J. N., Barnett, V., Coleman, K., & Johnston, A. E. (1994). Trends in herbage yields over the last century on the Rothamsted long-term continuous hay experiment. *Journal of Agricultural Science*, *122*, 365–365.

36 Coleman, S. Y., Shiel, R. S., & Evans, D. A. (1987). The effects of weather and nutrition on the yield of hay from Palace Leas meadow hay plots, at Cockle Park Experimental Farm, over the period from 1897 to 1980. *Grass and Forage Science*, *42*(4), 353–358.

37 Lynch, D., Henihan, A. M., Bowen, B., Lynch, D., McDonnell, K., Kwapinski, W., & Leahy, J. J. (2013). Utilisation of poultry litter as an energy feedstock. *Biomass and Bioenergy*, *49*, 197–204.

38 Öborn, I., Edwards, A. C., Witter, E., Oenema, O., Ivarsson, K., Withers, P. J. A., . Nilsson, S. I., & Richert Stinzing, A. (2003). Element balances as a tool for sustainable nutrient management: a critical appraisal of their merits and limitations within an agronomic and environmental context. *European Journal of Agronomy*, *20*(1), 211–225.

39 Lawes, J. B., & Gilbert, J. H. (1880). Agricultural, botanical, and chemical results of experiments on the mixed herbage of permanent meadow, conducted for more than twenty years in succession on the same land. Part I. *Philosophical Transactions of the Royal Society of London*, *171*, 289–416.

40 Lawes, J. B., & Gilbert, J. H. (1859). Experimental inquiry into the composition of some of the animals fed and slaughtered as human food. *Philosophical Transactions of the Royal Society of London*, *149*, 493–680.

41 Russeli, E. J. (1942). Rothamsted and its experiment station. *Agricultural History*, *16*(4), 161–183.

42 Sheail, J. (1996). Town wastes, agricultural sustainability and Victorian sewage. *Urban History*, *23*, 189–210.

43 Burt, T. P., Howden, N. J. K., Worrall, F., Whelan, M. J., & Bieroza, M. (2010). Nitrate in United Kingdom rivers: policy and its outcomes since 1970. *Environmental Science & Technology*, *45*(1), 175–181.

44 Addiscott, T. M. (1970). A note on resolving soil cation exchange capacity into mineral and organic fractions. *Journal of Agricultural Science, 75*, 365–367.

45 Scholz, R. W., Ulrich, A. E., Eilittä, M., & Roy, A. (2013). Sustainable use of phosphorus: a finite resource. *The Science of the Total Environment, 461*, 799–803.

46 Cooke, G. W. (1977). Waste of fertilizers. *Philosophical Transactions of the Royal Society of London. Series B, Biological Sciences, 281*(980), 231–241.

47 The Broadbalk Experiment at Rothamsted (2002). Film, produced by Vorst, J.J. and Martindale, W. http://www.agriculture.purdue.edu/broadbalk/ (accessed on 22 April 2014).

48 Coelli, T. J., & Rao, D. S. (2005). Total factor productivity growth in agriculture: a Malmquist index analysis of 93 countries, 1980–2000. *Agricultural Economics, 32*(s1), 115–134.

49 Lobell, D. B., Baldos, U. L. C., & Hertel, T. W. (2013). Climate adaptation as mitigation: the case of agricultural investments. *Environmental Research Letters, 8*(1), 015012.

50 Baker, J. S., Murray, B. C., McCarl, B. A., Feng, S., & Johansson, R. (2013). Implications of alternative agricultural productivity growth assumptions on land management, greenhouse gas emissions, and mitigation potential. *American Journal of Agricultural Economics, 95*(2), 435–441.

51 Burney, J. A., Davis, S. J., & Lobell, D. B. (2010). Greenhouse gas mitigation by agricultural intensification. *Proceedings of the National Academy of Sciences, 107*(26), 12052–12057.

52 Martindale, W., & Leegood, R. C. (1997). Acclimation of photosynthesis to low temperature in *Spinacia oleracea* L. II. Effects of nitrogen supply. *Journal of Experimental Botany, 48*(10), 1873–1880.

53 Grove, J. H., & Navarro, M. M. (2013). The problem is not N deficiency: active canopy sensors and chlorophyll meters detect P stress in corn and soybean. In J. V. Stafford (ed.), *Precision agriculture '13* (pp. 137–144).

Wageningen, The Netherlands: Wageningen Academic Publishers.

54 Sylvester-Bradley, R., & Kindred, D. R. (2009). Analysing nitrogen responses of cereals to prioritize routes to the improvement of nitrogen use efficiency. *Journal of Experimental Botany, 60*(7), 1939–1951.

3 The Scientific Basis for Food Security

3.1 The Supply of Essential Plant Nutrients

The conversion of solar energy into biomass through photosynthesis is dependent on several factors that can limit the amount of biomass or crop yield achieved. The notable limiting factors that growers and farmers are constantly trying to optimise for greatest crop yield are light, temperature, water, and plant nutrients. While light and temperature limitations are principally dealt with by sowing periods and protecting crops from weather, plant nutrients, and water are potential limiting factors that are perhaps most under the control of the management system the grower chooses. For example, limitations to water availability can be achieved by planning storage, capture, and irrigation of water resources, but ultimate availability may well be outside the control of growers. Plant nutrients are supplied as mineral fertilisers or organic manures, and these materials supplement nutrients available from the soil. What is critical to consider at this stage is that most essential nutrients required in foods by human beings are derived from biomass derived from crops and the agronomic decisions made during production will influence dietary quality.

Plant nutrients are found in crops at distinct concentrations that correspond to the ideal range for optimal plant health and the functioning of metabolism within the plant.

Global Food Security and Supply, First Edition. Wayne Martindale.
© 2015 John Wiley & Sons, Ltd. Published 2015 by John Wiley & Sons, Ltd.

It is possible to determine optimal physiological and metabolic states by the analysis of specific nutrients in field trials. There are critical values of plant nutrients below which there will be a decrease in yield, and the use of tissue analysis by agronomists is proving useful in determining critical nutrient values below which yield loss may become apparent. Critical values can also be used to manage crops at a particular level of production and determine optimal application of nutrients as fertilisers. The response curve of yield, metabolism, and physiological variables is generally hyperbolic, that is an initial linear phase of increase (of biomass yield) at low concentration of a variable such as nitrate, followed by a curvilinear phase at higher concentrations (Figure 3.1). This type of response can be viewed in terms of limiting factors and control theory, where several factors are involved in limiting the rate of response associated with a system. The mass of nutrient required per hectare of land is surprisingly

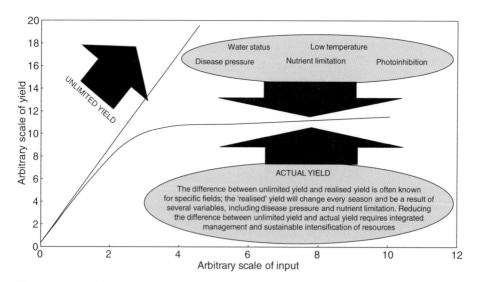

Figure 3.1. The law of limiting factors demonstrated using a case study of increasing material inputs that will increase the accumulation of biomass or crop yield. Martindale & Leegood, 1997.

stable across the plant kingdom and follows the type of responses shown in Figure 3.1.

This type of curvilinear relationship has been established since Lawes and Gilbert of Rothamsted developed the fertiliser recommendation for crops in the nineteenth century, it has continually been reviewed and revised for new technologies that have an impact on crop yield in agricultural systems.[1] It is currently viewed as a sum of several limiting factors that can be described in terms of control theory, that is, there are several control points in determining yield. Some control points may have more impact than others, and this value of impact will be dependent on environmental attributes, such as water availability, temperature, and light intensity.

Plant nutrients are classified on the basis of the mass required by plants to function metabolically and, in the case of crops, to provide optimal biomass quality and yield. The classification groups of the plant nutrients are described in Table 3.1. They include the major nutrients, the secondary nutrients, and the micro-nutrients or trace elements.

Table 3.1. The essential plant nutrients required for the production of all crop biomass

Nutrient (with common ion available to plants in brackets)	Symbol (fertiliser oxide declaration symbol in brackets)	Description and amount required by crops
		Macronutrients
Nitrogen (NO_3^-) Phosphorus (PO_4^{3-}) Potassium (K^+)	N P (P_2O_5) K (K_2O)	Between 40 and 400 kg nutrient required by a crop per hectare for optimal crop yield. Nitrate, very mobile; ammonium, not mobile; phosphate, not mobile; potassium, mobile

Table 3.1.   (Continued)

Nutrient (with common ion available to plants in brackets)	Symbol (fertiliser oxide declaration symbol in brackets)	Description and amount required by crops
		Minor nutrients
Sulfur (SO_3^{2-}) Calcium (Ca^{2+}) Magnesium (Mg^{2+})	S (SO3) Ca (CaO) Mg (MgO)	Typically 10–40 kg nutrient required by a crop per hectare for optimal yield. Sulphate, mobile; calcium, mobile; magnesium, mobile
		Micronutrients
Iron (Fe^{2+}) Boron Manganese Copper (Cu^{2+}) Zinc (Zn^{2+}) Molybdenum Chlorine (Cl^-)	Fe B Mn Cu Zn Mo Cl	Typically needed in quantities of 0.01–5 kg/ha Many micronutrients can be toxic at relatively low concentrations. The amount of micronutrient allowed in soils that have nutrients applied to them may be controlled by legislation. The availability of micronutrients is very much dependant on soil pH, with a general condition being greater availability of micronutrients and heavy metals at a pH below 7 in soils. This is not the condition for molybdenum, whose solubility increases at a pH above 7 in soils. Older soils >10 000 years old, very light soils are likely to have more trace element deficiencies than younger or heavier soils. Soil pH will be a critical factor in availability of trace elements.

Notes: As such, these nutrients maybe considered the base of the food supply chain. They are the inputs that are ultimately contained in foods that are added to or preserved by producers, processors and manufacturers.
Source: These data were adapted from the research of W. Martindale, published for Agronomy Extension Services as Part of the Fertiliser Advisers Certification and Training Scheme Technical Information Service, and Bøckman, Kaarstad, and Richards (1990).[2]

The nutritional balance between the agricultural biomass producer to the food processor and manufacture in the supply chain is not only about the interface or input–output ratio of the 13 essential plant nutrients described in Table 3.1. The essential plant nutrients can optimise plant metabolism for production and influence the nutritional quality of biomass by their impact on the production of an important group of metabolites called phytonutrients that include pigments, flavours, fragrances, metabolic co-factors, and vitamins.

3.2   Plant Nutrients and Phytonutrients in the Food Supply Chain: Establishing a Nutritional Understanding Using Human Trials

Phytonutrients are a group of nutritional chemicals that are grouped collectively because they are of high nutritional value whose concentration in crop biomass is drastically changed by crop variety, agronomic management, and preservation in the supply chain.[3] The phytonutrients include familiar dietary nutrients, such as vitamin A, vitamin C, vitamin B, vitamin E, carotenoids, lycopenes, and flavonoids. These are often collectively described as antioxidants in that they reduce the activity of potentially harmful free radicals in cells that result from metabolic reactions. This function of 'mopping-up' of oxidising capacity is linked to providing anti-carcinogenic impact, enhancing healthy metabolic attributes and improving several specific organ functions. The existence of data to support the functions of phytonutrients and antioxidants in the diets of populations is often limited by the number of people included in nutritional trials, but evidence continues to increase.[4] There are excellent examples of nutritional research on micronutrient and protein consumption that can be integrated into diet and meal planning so the link between phytonutrient consumption and health can be realised by consumers.

This practice is essential if we are to enact and enable the application of nutritional research for consumers. Indeed, it may be seen as the equivalent of extension practice in agriculture for management techniques because it is delivering products with very clear nutritional aims for the supply chain. An impressive example of applying research from nutritional trials to consumer requirements is the CSIRO Total Well Being Diet, where studies have determined optimal protein intake for healthy lifestyles.[5] This diet is available to consumers as a book, and it communicates CSIRO's protein consumption research, CSIRO is the Australian Government science agency and this book has sold in excess of 700 000 copies. This diet shows there is significant interest in changing diets for healthier and more sustainable lifestyles, and there are routes to make this more likely that use nutritional trials to demonstrate outcomes. A similar approach has been used by the World Wide Fund for Nature (WWF) to promote not only healthy diet planning but increased sustainability criteria associated with healthier meals.[6]

At a food system world view, the micronutrients and phytonutrients in food supply chains are of critical importance because of the emergence of what are known as 'hidden hungers'. These have become important in policy arenas that aim to increase food security by enhancing the value of meal planning and health outcomes associated with different diets. This is not the food security associated with the consumption of energy and protein alone, it is concerned with the balance or quality of foods consumed. Indeed, to many commentators, overlooking the 'hidden hungers' has been responsible for poor dietary policies in both economically undeveloped and developed nations. This is because the primary responses of policymakers who wish to reach food security are focused on the supply of calories and protein without a considering micronutrition impacts. Indeed, global statistics and data concerned with the production and trade of primary agricultural products and foods do not report phytonutrient quality of these materials. This lack of insight can ultimately lead to deficiencies in micronutrients that have been suggested to result in the overconsumption of calories and protein.[7] The micronutrients for which there are well documented cases of 'hidden-hunger' impacts include iron and the relationship of intake and bioaccumulation with medical conditions such as anaemia and cognitive development.[8]

A well-documented hidden-hunger scenario is also provided by selenium, which is essential to human and animal health and not for crop growth. It is therefore not an essential plant nutrient whose accumulation in the supply chain is critical because it is an important human nutrient. Reported research suggests that deficiency in selenium can result in over consumption of foods because more food is eaten to obtain necessary selenium and other micronutrient intake. This type of hidden-hunger scenario is thought possible for most micronutrients and phytonutrients, and it has developed innovative views of what sustainable dietary advice could be.[9]

Thus, attention is placed on the mineral composition of agri-produce and the hidden-hunger scenarios because zinc, selenium, copper, and perhaps most globally important, iron content of crop biomass, is influenced by soils and the plant nutrients applied to crops. These micronutrients do vary in the composition of cereal produce around the world and reflect the fertilisation policy and soil composition to a large extent.[10] The optimal mix of foliage, seed, and root tissues can also influence micronutrient intake, as can how foods are preserved or cooked. These attributes of efficient micronutrient delivery are continuously overlooked in nutritional policy, and the benefits of using biofortified crop varieties, such as golden rice have yet to be realised.[11]

The phytonutrient content of biomass is dependent on crop species and crop variety, and specific crops are often grown for particular phytonutrients for human consumption. Indeed, the development of different-coloured vegetable and fruit varieties is associated with differing phtyonutrient content. The selection of different foods for meals based on different colour and fragrance stimuli is an important route to consuming a healthy balance of phytonutrients. This approach of developing crop variety for specific nutritional properties has led to the production of crops for pharmaceutical and industrial feedstocks. An example is the development of increased phyto-oestrogen levels in the soybean crop; these are most notably present in soya extracts and are thought important for a number of metabolic functions including maintaining bone health.[12] Other compounds include the brassinosteroids found in crucifers that are suggested to offers anti-carcinogenic benefits when they are consumed.[13] Perhaps the most lauded phytonutrients of a specialised nature are the poly-unsaturated fatty acids (PUFAs) and mono-unsaturated fatty acids (MUFAs) because the fatty acid composition of agri-produce has become extremely important, with linola, borage, and meadowfoam oils being grown for health supplements. The production of oils that contain increased levels of MUFAs and PUFAs has become more important for reasons relating to the health of the human circulatory system. The oilseed crops that contain the omega unsaturated fatty acids at increased levels include nuts, such as walnut and hemp, and these oils are found in oily fish and algae. The associated health benefits of increased MUFA and PUFA diets were first recognised in populations that had increased consumption of oily fish products.[14,15]

Naturally, these types of nutrients are contained within the biomass that is produced by agricultural systems, and their concentration in biomass can be enhanced by using plant nutrients applied to crops as fertiliser, which is known as agrifortification. There is also a requirement for efficient supply chains to optimize the preservation of food nutritional value or add to it by fortifying biomass and food materials in the food chain. The agrifortification and biofortifcation of foods is discussed here in the context of enhancing the delivery of phytonutrients in diet, but we do need an understanding of how nutrients are accumulated in biomass within the agricultural system. The interaction between plant genes that result in the production of phytonutrients and human health will provide many future opportunities in bringing diets to consumers that are more focused to individuals' health requirements. The role of genetically engineering crops to provide these specific nutrients offers huge potential.[16]

3.3 Biomass, the Base of the Supply Chain

Understanding the principles of biomass production is crucial to understanding the nutritional interface between production and food manufacture because it not only provides calories and protein, but also the materials and cellular components in which micronutrients and phytonutrients are contained. This requires a thorough understanding of how crop nutrition and crop protection are used in farming systems around the world because the use of these inputs stimulate the production of biomass and interception of solar energy. Thus, agronomic practices will influence the quality of both plant and animal biomass produced from the farming system. The principles of producing biomass in the farming system for quantity and quality of food supply chain products can be placed in the context of what are called integrated management regimes, where all aspects of crop production are considered within the food system. This includes from the molecular scale of biochemistry and physiology through to the macroscale geography and land use planning impacts that are utilised to reach optimal management decisions.

Integrated management techniques are an important suite of knowledge and thinking used by producers in the food supply chain and they should be utilised by manufacturers and processors in future. We have already introduced the value of Integrated Plant Nutrition Systems here where integrated management brings together established routes for the optimal management of producing biomass for food supply and integrates them with emergent technologies and policy frameworks. The branches of integrated management include the following:

1. **Integrated Crop Management (ICM):** The principles of ICM are associated with determining optimal energy inputs for crop yield that provide economic and environmental outcomes that enable sustainable business practice.[17]
2. **Integrated Pest Management (IPM);** The principles of IPM aim to optimise the yield of crops and livestock by reducing the costs of pest control and the impact of disease. The use of scheduled pest control actions in response to observed pest thresholds are the focus of IPM.[18]
3. **Integrated Plant Nutrient Systems (IPNS):** The principles of IPNS aim to optimise crop and livestock yield in response to plant nutrient management on farms. The use of nutrient management plans, nutrient balances, and fertiliser recommendations are the focus of IPNS.[19]

The integrated approach provides a system-thinking view of farming operations in that no single operation used to optimise yield is taken in isolation and the goal of sustainable yield is made possible by considering all data available to growers and farmers. An important consideration already mentioned that is often outside of the direct control of the producer is that of climate, where the interception of light, impact of ambient temperature, and acquisition of water have impacts on biomass yield. While out of direct control, understanding how these resources are used can point producers to potential areas of efficiency improvement within the agricultural system.

3.4   The Interception of Light by Crop Canopies: How the Molecular Scale Impacts on Food Supply Chain Efficiency

The interception of solar radiation will be dependent on the leaf area intercepting energy from sunlight, as the leaf area index (LAI) shows; although canopy structures and effects of shading do make this statement very simplistic, its importance is often overlooked. LAI is the area of leaf per unit ground cover, and the optimisation of LAI has been crucial in developing modern wheat varieties.[20] The duration a leaf is expanded and effectively photosynthesising is also important in determining the amount of energy intercepted, and this is measured as the leaf area duration (LAD). There are very good correlations between light interception, dry matter increase, and LAI and LAD. Crops will increase LAI and LAD to a point where metabolism can be maintained, and surplus energy intercepted is diverted into dry matter gain; optimising this process through integrated management or crop breeding is clearly important to increasing crop yield. While this is a relatively simple explanation that considers many different environmental variables that determine leaf expansion and duration, the largest leaves of crops are proven to be important indicators of potential yield capacity.[21] What research has achieved by understanding the interaction with LAI and LAD with yields is to make what could be just anecdotal observations in the field into a specific metabolic outcomes that relates to crop performance and eventual biomass yield. Measuring the effectiveness of light capture by canopies is an important indicator for crop yield, and the activity of leaf photosystems that convert solar energy into dry matter, understanding and measuring how the activity of these photosystems changes dry matter accumulation, has important implications for projecting biomass yield.

Global food production is dependent on the primary dry matter produced from capturing solar energy by expanded leaves. Now that we have an understanding of both the nutrient balance of agriculture and how plant nutrients can be managed, we should consider how crops produce biomass from these nutrients. The key processes to understand are what might be regarded as a suite of biochemical reactions that utilise solar energy to produce both carbohydrates and energy-transferring molecules that enable the synthesis of proteins, plant growth, and the accumulation of biomass. These energy transfers from solar to chemical energy sources are described by the metabolic reactions of photosynthesis, and the outcomes of these reactions are subject to the laws of thermodynamics. This is because photosynthesis is essentially an energy transferring activity where solar energy and heat is captured, which is not 100% efficient because there are many transfer and capture points where energy is dissipated as heat or entropy. This is an outcome of the second law of thermodynamics, which states energy can be transferred within a system but in doing so disorder or entropy is dissipated. At a macro- or physiological-scale energy transfer, inefficiencies are identified by the crop canopy interception of photosynthetically active radiation (PAR), where less than 5% PAR is typically intercepted by even the most efficient crop canopy. This is because PAR is made up of only the red and blue parts of the visible spectrum of light, and PAR must be captured by photosystems in the leaf; there are many different processes involved in getting the right type of energy to the right place—it cannot be a 100% efficient process.

At a molecular scale, the production of dry matter or biomass in plants is determined by the biochemical reactions that result in the synthesis and transport of carbohydrates that are produced from photosynthesis. The solar energy captured by crop leaf canopies is used by a number of relatively conserved chemical energy shunts that utilise the energy of electron transfer occurring in Light Harvesting Complexess (LHCs) within leaves that include pigments, such as chlorophyll. The electron transfer processes occurring in the LHCs are powered by the capture of PAR in crop canopies. This electron shunting results in the release of energy that is captured by soluble macromolecules, such as adenosine tri-phosphate (ATP) and nicotine nucleotides. While the biochemical processes described here are simplified for purposes of explanation, they have often been considered abstract when we consider the macroscale of field agronomy. However, understanding how photosynthesis is regulated by metabolic factors within leaves and environmental changes outside leaves is crucial to any agricultural management system, and it offers much future opportunity in stimulating management changes and enabling the development of improved crop varieties.

Indeed, this understanding of how molecular knowledge links to agronomic and food supply chain efficiency is the basis for the development of the bioengineering and biotechnology industries. The application of biotechnology can change both the macroscale factors, such as leaf expansion, and microscale factors, such as the transport of carbohydrate, that are influenced by the transfer of solar to chemical energy. Biotechnology has enabled the sequencing of crop genomes so that metabolic events associated with yield can be understood with respect to the regulation of specific genes. Furthermore, the establishment of genome knowledge has identified groups of genes that determine macroscale changes in crops such as leaf expansion, nutrient accumulation, and senescence or leaf duration.[22] The metabolic responses to changes in the environmental variables, such as light intensity and temperature, also provide signals that can be detected by sensors, such as chlorophyll meters, that are used to determine crop health. Thus, the activity of specific genes or groups of genes that can be used to identify stress brought about by metabolic changes provides a type of sensing that can improve agronomic management.[23]

Photosynthesis is carried out by catalytic proteins or enzymes in plant leaves and other green tissues. The soluble enzymatic proteins that are not part of the photosystem proteins associated with membranes are functional proteins that enable the biochemical reactions that reduce carbon dioxide to carbohydrate, which in turn may accumulate to increase the dry matter in a crop. Other proteins have storage functions and accumulate in storage tissues, such as seeds; together with the accumulation of carbohydrates, the functional and storage protein content of crops often determine their nutritive value. The taste and texture attributes associated with crop ingredients are determined by many other metabolites, such as pigments and essential oils, but protein content is a good indicator of nutritional value. Many other metabolic processes will determine the eventual dry matter accumulation, including the 'storage capacity' of a crop to hold carbohydrate and its ability to transport carbohydrate to storage tissues. In many cases, crops have been bred by recognising these capture, transport, and storage traits, which are often observable in the field, to select for increased storage capacity for carbohydrate. Thus, the ability to observe what is known as 'sink strength' has been critical to the development of modern crop varieties because many of the storage tissues we harvest, such as seeds, tubers, and roots, are the result of selecting for increased sink strength and storage capacity which can be measured by field observation.

We are now able to define sink strength activities more precisely by using knowledge of the role of genes and groups of genes that control them. New biotechnological techniques in genomics can identify genes that control sink strength and provide targets for selection to improve performance in the field. An understanding of how the protein and carbohydrate content of plants are regulated in agricultural practice is critical if we are to breed more productive crops and are to understand the eventual nutritional qualities of foods and food stuffs.

3.5 The Requirement for Breeding New Crop Varieties and Selecting for Increased Sink Capacity of Crops

We know that sink strength is an extremely important component of yield determination and a complex parameter to quantify, and there are demonstrations of the ability of sink strength to influence yield and dry matter production. These are provided by greenhouse gas and global warming experiments, where plants are grown at increased CO_2 concentrations. Acclimation to increased CO_2 results in a decreased photosynthetic capacity over a longer term period. Acclimation results from changes in gene expression over days and not changes in metabolite concentration or enzyme activation that occur over minutes in response to environmental change. When photosynthetic tissue is harvested and allowed to regrow, such as for forage grasses, they experience a increase in sink strength even at increased CO_2 concentrations and photosynthetic capacity increases. It is as if the photosynthetic system works harder at increased CO_2 and the plant becomes clogged with assimilates, such as sucrose and starch; when the clogged sinks are removed, sink capacity increases again. Nitrogen can ameliorate this effect because plants grown with increased nitrogen continue to increase photosynthetic capacity at increased CO_2; this is because sufficient nitrogen is available to develop new tissues and maintain sink strength. At increased CO_2 and limiting nitrogen conditions, the acclimatory response of decreased photosynthetic capacity is amplified and more apparent than at high nitrogen availability. Thus, global warming and environmental change could have important implications for crop yield and nitrogen requirements of the future food system.

3.6 Photosynthetic Metabolism, the Biochemical Driver of Production

We now understand that photosynthesis requires proteins to catalyse reactions, and a typical protein contains 5% nitrogen and 1% sulfur by weight so they are substantial sinks for the nitrogen and sulfur nutrients in crops. These proteins, together with storage proteins, are the source of nutritional variation in many foods. The development of crops that preferentially store protein or carbohydrate in sink or storage tissue is the basis for growing many of the crops we utilise for protein or carbohydrate sources in food processing and manufacture. The protein investment made in photosynthesis is large, with a single photosynthetic protein, the Rubisco protein, making up 25% of a typical leaf protein quota. Rubisco is an acronym for ribulose-1,5-bisphosphate carboxylase-oxygenase; it is a soluble protein contained in chloroplasts within photosynthetic cells that fixes carbon dioxide gas into a reduced carbohydrate form that is synthesised into what we call sugar or sucrose, the white crystalline sugar utilised by the food system.

Sucrose is the universal carbohydrate currency of all plants, and it is a disaccharide of glucose and fructose that can be metabolised to produce transferrable energy or stored as a glucose polymer as starch. But it is the protein allocation of photosynthesis that concerns us here because the concentration of Rubisco protein in a leaf may provide a means of assessing the productivity of a crop by again acting as a sensor or indicator for production of dry matter. A further substantial protein pool within the leaf is the insoluble membrane bound or insoluble proteins within the chloroplast that are associated with light harvesting complex (LHC) or photosystem functions and pigments, such as chlorophyll and carotene. This group of proteins can make up 40% of the leaf protein total. The total photosystem insoluble protein and soluble enzyme protein in a typical leaf can account for at least 70% of the protein of a plant, and this represents a major requirement for nitrogen, sulfur, and other essential plant nutrients. Thus, the protein allocation and physiological responses of photosynthesis are important targets for methods that will help agriculturalists manage crops for the food supply chain and provide sensors for productivity.[24] What is crucial in all of this is that we not only account for quantity but also for quality when we consider crop yields.

The relationship between nitrogen and leaf biochemistry demonstrates that nitrogen can substantially limit production, but above a critical leaf concentration of nitrogen, the benefits of obtaining more nitrogen are negligible. We have already discussed this with respect to increased atmospheric carbon dioxide concentration and acclimatory responses of crops to environmental stresses. The acclimatory responses of plants have implications for how nitrogen is mobilised in canopies, how much nitrogen needs to be supplied to a crop, and how much is stored in harvested biomass. The interaction between nitrogen use and distribution within the canopy cover and architecture of plants offers a means of utilising crops as sensors in optimal production systems. The basic metabolic responses of crops will be generic across most cultivated species; however, subtle but important differences exist that can provide specific targets for crop improvement. The following examples of metabolic targets demonstrate how biotechnologies can utilise natural systems to manipulate yield and productivity in agricultural systems:

1. Water stress is a target for agronomic improvement because crop varieties that are able to tolerate increased

levels of water stress will enable the most efficient use of land experiencing climate change[25-27]

2. Nitrogen responders are those crops that can either provide sensors for specific nutrient requirements (in this case nitrogen) or those crops that will accumulate dry matter most effectively in response to nitrogen. They are of agronomic value because of the strong relationship between nitrogen use efficiency and crop yield.[28]

3. Pest control using biotechnological techniques, such as inducing crops to produce pheromones that deter insect pests or develop metabolic responses that reduce the impact of diseases offer new routes to deploying crop protection strategies.[29,30]

4. Nutrient composition of crops has been identified as an area of improvement opportunities for biofortifying crops, and with the use of biotechnologies, nutritional improvement of crops will become more innovative. This is apparent with respect to now being able to design crops with specific nutrient profiles such as those having different oil composition in their seeds.[31]

3.7 Environmental Stress Events and Their Impacts on Food Supply

An important area where metabolism, environmental stressors, and biotechnological opportunities exist is the understanding of how crops respond to changes in light and temperature. The combination of these two environmental variables can result in decreased temperature, which slows metabolism, and increased light intensity, which increases photoreactions of photosynthesis. These low temperature and high light conditions are not uncommon in field situations, and they result in photoinhibition in plants. In effect, plants have developed sun protection by modifying pigment concentrations in the LHCs and membranes structures to dissipate excess solar energy as heat and protect metabolic processes from photoinhibition. Photoinhibition results in the production of previously mentioned free radicals whose damaging activities are reduced by specific antioxidants that are included in the phytonutrient group of metabolites we have already discussed.

The mechanisms of nutrient partitioning have been investigated since the 1960s with regard to leaf and canopy structures, and the biochemical elucidation of photosynthesis has enabled an understanding of carbohydrate synthesis in low temperature and high light environments to be understood in the 1980s. Biotechnological techniques enabled manipulation of pigment genes that conferred elements of low temperature resistance and increased acclimation in the 1990s. The development of low temperature and frost resistance in crops is of clear agronomic application since susceptibility to low temperature during germination and grain filling can have detrimental impact on yield. Indeed, the following case study shows how biotechnological techniques have enabled, firstly, understanding low temperature responses of crops, and then to target specific metabolic pathways that enable crops to acclimate or protect cells from low and freezing temperatures.[32]

Low temperature decreases photosynthetic capacity, and this can result in the acclimation of crops to lower temperatures. Acclimation will result in an increase in the ability of leaves to carry out gas exchange, that is, carbon dioxide entering and water vapour being removed from the leaves via the stomata. The response to decreased temperature will be associated with a number of variables, including the acclimation of stomata and many metabolic changes. These responses can be limited by protein synthesis, and this can be demonstrated using the photosynthetic response curves that show a limiting factor response we have seen previously. These show that acclimation to low temperature and high light is associated with an increase in the rate of carbon assimilation and the capacity to utilise photosynthetic metabolites in cells. An increase in the capacity of gas exchange of the leaf during photosynthesis is related to increased amounts of photosynthetic enzymes within the leaf. An important enzyme is Rubisco, which is a sensor for crop health and production capacity; it makes up between 20% and 30% of the leaf soluble protein and therefore is easy to detect and its activity is easily measured. Placing these diagnostic methods from laboratories into field situations is challenging, but the detection of the amount of Rubisco in leaves or greenness of leaves may provide a means of determining the health and acclimation potential of crops. An important implication of acclimation to low temperature is that increases in enzyme activities will require a greater amount of protein synthesis, and therefore nitrogen availability must be increased or nitrogen use efficiency improved.[33] This is similarly the case for phosphate use in photosynthetic cells, since the amount of sugar phosphate metabolites in cold-acclimated leaves is known to increase. An increase in the capacity to synthesis soluble sugars, principally sucrose in many plants, is closely related to the ability for crops to acclimate to low temperature where sugars such as sucrose can act as cryoprotectants that maintain the integrity of cell membranes.[34] Thus, a consequence of low temperature acclimation is the production of free sugars such as sucrose in cells that confer cryoprotective effects on membranes within cells. High concentrations of sucrose increase the viscosity of fluid around membranes, thus protecting them against the disruptive formation of ice crystals. This principle is used in food preservation and freezing, where the integrity of foods can be maintained during cold treatment by maintaining free sugar concentrations. Yet again, this type of process can be influenced by agronomic management of biomass in that foliage can be low temperature sweetened prior to freezing so that the concentration of free sugars is increased.

This has implications for plant storage organs that are cold stored, such as potatoes, sugar beet, and many fruit and vegetable crops. The storage and low temperature sweetening of potatoes, fruits, and vegetables for the retail market is a recognised characteristic associated with changes in taste. There has been substantial interest in finding out what determines cold sweetening in potatoes that undergo storage periods at low temperature. Sweetening has implications for the processing industry because free reducing sugars in potato tissue can cause discolouration during frying of potato chips. Increased sugar synthesis in plants at low temperature may also provide insights into the improved storage of phytonutrients since specific sugars, such as trehalose, synthesised in plants and microbes have extremely good cryoprotective properties.

The light environment of leaves can stimulate acclimatory responses of the light capture systems and other photosynthetic components of a leaf if the light environment is decreased or increased in intensity within a canopy. Under such conditions, photoacclimatory responses occur to maintain photosynthetic capacity, and if acclimation does not occur, photosystems in leaves become super energised with no sink for absorbed PAR and energy, resulting in increased photoinhibiton processes within a leaf, which can decrease productivity. Light is rarely thought of limiting photosynthesis when it is at high intensities, although photoinhibition has been shown to occur in all plant species to some degree at all times if photosystems are operating at maximum rates. There are metabolic mechanisms that result from photoinhibition of the light capturing systems in leaves that cause the degradation of proteins involved in electron transport, thus decreasing energy capture and photosynthetic production. Furthermore, there is a metabolic mechanism known as photorespiration that results in the consumption of oxygen and results in the release of CO_2 and an ammonia molecule. This process of photorespiration is thought to be important in dissipating light energy at high light intensities. Many temperate crops will experience photoinhibitory conditions of high light and low temperature, particularly if they are autumn sown. Hence, understanding the relationship between photosynthesis, photoinhibition, and photorespiration with environmental variables can help us understand how crop metabolism responds to these particular stresses. These types of investigation may also identify metabolic indicators of stress that can be utilised by agronomists.

3.8   The Principles of Integrated Management across the Food Chain: A Food Supply Chain Perspective

Integrated management of pests, nutrients, and crops depends on a number of different approaches to optimising yield being utilised to develop profitable and sustainable farms. They can include the following headline actions.

1. The use of Geographic Information Systems (GIS's) has started to replace traditional field maps and have the potential to revolutionise farming practices. GIS have provided more fit-for-purpose means of using maps within an Information Technology (IT) framework; the map remains the same whether it is digital or on paper or transparency formats. Indeed, this is the principle of precision agriculture and the techniques associated with it that target agricultural inputs and outputs to specific areas. The use of soil and plant sensors to target inputs is becoming important in farming systems, and metabolic sensors have been discussed. The use of precision agricultural techniques for outputs has focused on harvesting efficiency and transport of products. All farm maps will include general reference points, including streams, residences, well heads, number of hectares, and soil types. This is the basis for the rest of the plan. The use of GIS systems to map farms is

becoming more common as computer power increases and their price decreases. Maps of sufficient resolution for farms are available free or at minimal cost on the World Wide Web and 'off-the-shelf' software providing GIS applications. GIS software has become compatible with other standard software packages that handle spreadsheets and databases so that a GIS system can be interfaced with other expert systems. These systems are able to store long-term records for assurance and traceability purposes. Furthermore, GIS systems can be utilised to map activities across food supply chains from producers to consumer functions.

2. Soil and agricultural product testing is critical because this will determine how much nutrient is being utilised as N, P, K, trace nutrients and provide assessment of pH and SOM so that nutrient balances can be developed. Again, the use of analytical data provides a means to use sensor-related and threshold values for developing management actions, such as the use of fertiliser, pesticides, and processing operations.

3. Cropping plans should be prepared to represent market trends and the environmental sustainability of farming enterprises. This will include assessment of estimated yield; factors affecting yield are numerous and complex. Using historic series of yield records is important in developing yield estimates, and the more accurate the yield estimate, the easier it is to potentially optimise input use efficiency. Furthermore, the use of market information regarding the price trends and volatility agricultural products and food ingredients will be utilised in determining crop plans.

3.9 The Modern Agricultural System, the Dietary Interface, and Food Supply

Modern agriculture exists in a rapidly changing commercial, regulatory, and economic environment where new technologies, genetically modified crops, food traceability and safety, environmental quality, and financial survival in global markets must be balanced within businesses. The only way to achieve this is by considering the farm enterprise as a system where technologies enable the production of biomass, improve crop and livestock production, and result in improved environmental management. The food production systems of the world depend upon primary production of dry matter or biomass by plants to provide ingredients for the manufacturing and processing functions of the food supply chain. Biomass and crop products are consumed in our food chain directly by us, or via livestock as meat and dairy produce or recycled within the biosphere as an energy source. The integration of biological knowledge into understanding plant production systems is an important goal for growers of biomass to strive for. The historical development of farming systems shows us that extension and agri-education are an important part of disseminating management knowledge, particularly when they are concerned with new technologies and methods of utilising food ingredients. Extension of biological principles to the food supply chain requires a framework of analysis, such as LCA or energy balance to incorporate economic, environmental, and agronomic decisions together so that they can be measured and assessed in terms of their performance.

The basis for any decision made in the food supply chain made by the farmer, manufacturer, retailer, or consumer must integrate economic, social, and environmental considerations. Here is our first and foremost difficulty in understanding the balance of both quantitative and qualitative attributes that result in management decisions. Ultimately, the decisions made will be an integration of several sources and disciplines. What a formalised integrative approach, such as LCA and energy balance, does is to provide terms of reference that performance can be measured against. This provides a platform to explain decisions and evaluate them in increasingly important arenas, such as the global trade of products, where supply chains are held to account by regulators for actions around the world that need to be reported to achieve optimal profitability, communicate responsibility, and ensure food safety.

Interfacing the food supply chain with the initial points of field trials and agricultural operations is critical because we have seen that how crops are grown drastically influences food quantity and quality. It is possible to detect metabolic changes in crops at field and landscape scales so that these foods-based quality attributes can be managed. This means precision techniques that map inputs and outputs from cultivation to manufacturing and processing are valuable tools that will tackle threats to yield loss and enhance the ability to develop management systems that can meet the demands of the food supply chain.

References

1 Scarseth, G. D. (1962). *Man and his earth.* Ames: Iowa State University Press.
2 Bøckman, O. C., Kaarstad, O., Lie, O. H., & Richards, I. (1990). *Agriculture and fertilizers.* CAB Direct Record Number 19901948903. Oslo: Norsk Hydro AS.
3 Martin, C., Butelli, E., Petroni, K., & Tonelli, C. (2011). How can research on plants contribute to promoting human health? *The Plant Cell Online, 23*(5), 1685–1699.
4 Ames, B. N. (2004). A role for supplements in optimizing health: the metabolic tune-up. *Archives of Biochemistry and Biophysics, 423*(1), 227–234.
5 Noakes, M., & Clifton, P. M. (2006). *The CSIRO total well-being diet.* Camberwell, Australia: Penguin.
6 Macdiarmid, J., Kyle, J., Horgan, G., Loe, J., Fyfe, C., Johnstone, A., & McNeill, G. (2011). Livewell: a balance of healthy and sustainable food choices. Livewell report 2011. London: WWF UK.

7   Ames, B. N. (2006). Low micronutrient intake may accelerate the degenerative diseases of aging through allocation of scarce micronutrients by triage. *Proceedings of the National Academy of Sciences, 103*(47), 17589–17594.

8   McCann, J. C., & Ames, B. N. (2007). An overview of evidence for a causal relation between iron deficiency during development and deficits in cognitive or behavioral function. *The American Journal of Clinical Nutrition, 85*(4), 931–945.

9   Ames, B. N. (2006). Low micronutrient intake may accelerate the degenerative diseases of aging through allocation of scarce micronutrients by triage. *Proceedings of the National Academy of Sciences, 103*(47), 17589–17594.

10 Welch, R. M., & Graham, R. D. (2004). Breeding for micronutrients in staple food crops from a human nutrition perspective. *Journal of Experimental Botany, 55*(396), 353–364.

11 Martin, C., Butelli, E., Petroni, K., & Tonelli, C. (2011). How can research on plants contribute to promoting human health? *The Plant Cell Online, 23*(5), 1685–1699.

12 Branca, F. (2003, November). Dietary phyto-oestrogens and bone health. *Proceedings: Nutrition Society of London, 62*(4), 877–887). CABI Publishing. 1999.

13 Moreau, R. A., Whitaker, B. D., & Hicks, K. B. (2002). Phytosterols, phytostanols, and their conjugates in foods: structural diversity, quantitative analysis, and health-promoting uses. *Progress in Lipid Research, 41*(6), 457–500.

14 McCann, J. C., & Ames, B. N. (2005). Is docosahexaenoic acid, an n–3 long-chain polyunsaturated fatty acid, required for development of normal brain function? An overview of evidence from cognitive and behavioral tests in humans and animals. *The American Journal of Clinical Nutrition, 82*(2), 281–295.

15 Kris-Etherton, P. M., Harris, W. S., & Appel, L. J. (2003). Fish consumption, fish oil, omega-3 fatty acids, and cardiovascular disease. *Arteriosclerosis, Thrombosis, and Vascular Biology, 23*(2), e20–e30.

16 Martin, C. (2012). The interface between plant metabolic engineering and human health. *Current Opinion in Biotechnology, 24*(2), 344–353.

17 Leake, A. (2000). The development of integrated crop management in agricultural crops: comparisons with conventional methods. *Pest Management Science, 56*(11), 950–953.

18 Barzman, M., & Dachbrodt-Saaydeh, S. (2011). Comparative analysis of pesticide action plans in five European countries. *Pest Management Science, 67*(12), 1481–1485.

19 Dicks, L. V., Bardgett, R. D., Bell, J., Benton, T. G., Booth, A., Bouwman, J., Brown, C., Bruce, A., Burgess, P. J., Butler, S. J., Crute, I., Dixon, F., Drummond, C., Freckleton, R. P., Gill, M., Graham, A., Hails, R. S., Hallett, J., Hart, B., Hillier, J. G., Holland, J. M., Huxley, J. N., Ingram, J. S. I., King, V., MacMillan, T., McGonigle, D.F., McQuaid, C., Nevard, T., Norman, S., Norris, K., Pazderka, C., Poonaji, I., Quinn, C. H., Ramsden, S. J., Sinclair, D., Siriwardena, G. M., Vickery, J.A., Whitmore, A. P., Wolmer, W. & Sutherland, W. J. (2013). What do we need to know to enhance the environmental sustainability of agricultural production? A prioritisation of knowledge needs for the UK food system. *Sustainability, 5*(7), 3095–3115.

20 Austin, R. B. (1999). Yield of wheat in the United Kingdom: recent advances and prospects. *Crop Science, 39*(6), 1604–1610.

21 Evans, L. T., & Fischer, R. A. (1999). Yield potential: its definition, measurement, and significance. *Crop Science, 39*(6), 1544–1551.

22 Varshney, R. K., Hoisington, D. A., & Tyagi, A. K. (2006). Advances in cereal genomics and applications in crop breeding. *Trends in Biotechnology, 24*(11), 490–499.

23 Varshney, R. K., Graner, A., & Sorrells, M. E. (2005). Genomics-assisted breeding for crop improvement. *Trends in Plant Science, 10*(12), 621–630.

24 Pirie, N. W. (ed.) (2012). *Food protein sources* (Vol. 4). Cambridge, UK: Cambridge University Press.

25 Roy, S. J., Tucker, E. J., & Tester, M. (2011). Genetic analysis of abiotic stress tolerance in crops. *Current Opinion in Plant Biology, 14*(3), 232–239.

26 Eisenstein, M. (2013). Plant breeding: discovery in a dry spell. *Nature, 501*(7468), S7–S9.

27 Peleg, Z., Apse, M. P., & Blumwald, E. (2011). Engineering salinity and water-stress tolerance in crop plants: getting closer to the field. *Advances in Botanical Research*, *57*, 405–443.

28 Schachtman, D. P. (2012). Recent advances in nutrient sensing and signaling. *Molecular Plant*, *5*(6), 1170–1172.

29 Gross, M. (2011). New directions in crop protection. *Current Biology*, *21*(17), R641–R643.

30 Aliferis, K. A., & Jabaji, S. (2011). Metabolomics—a robust bioanalytical approach for the discovery of the modes-of-action of pesticides: a review. *Pesticide Biochemistry and Physiology*, *100*(2), 105–117.

31 Haslam, R. P., Ruiz-Lopez, N., Eastmond, P., Moloney, M., Sayanova, O., & Napier, J. A. (2013). The modification of plant oil composition via metabolic engineering: better nutrition by design. *Plant Biotechnology Journal*, *11*(2), 157–168.

32 Guy, C. L. (1990). Cold acclimation and freezing stress tolerance: role of protein metabolism. *Annual Review of Plant Biology*, *41*(1), 187–223.

33 Holaday, A. S., Martindale, W., Alred, R., Brooks, A. L., & Leegood, R. C. (1992). Changes in activities of enzymes of carbon metabolism in leaves during exposure of plants to low temperature. *Plant Physiology*, *98*(3), 1105–1114.

34 Martindale, W., & Leegood, R. C. (1997). Acclimation of photosynthesis to low temperature in *Spinacia oleracea* L. I. Effects of acclimation on CO2 assimilation and carbon partitioning. *Journal of Experimental Botany*, *48*(10), 1865–1872.

4 The Sociological Basis for Food Security

4.1 Challenges and Solutions

Using the analysis of the food system and food supply chains developed in the previous chapters, we have established what food security is and why we should care about it. The following three central themes have emerged as presenting the greatest challenges to the food supply chain:

1. The increasing reliance on food trade and trade agreements to develop food security means that socio-political attributes of increasing the investment and infrastructural drivers are just as important as maintaining innovations associated with new technologies entering supply chains.
2. Using localised food consumption to provide sustainable solutions to food security will enhance consumer knowledge of foods and qualities of foods.
3. The requirement to deal with the finite nature of the resources we use to produce food is crucial at all stages of the food supply chain. Minimising waste is something that must happen; it represents some of the greatst challenges facing the food security debate and can be viewed as 'the elephant in the room' we can no longer ignore.

Global Food Security and Supply, First Edition. Wayne Martindale.
© 2015 John Wiley & Sons, Ltd. Published 2015 by John Wiley & Sons, Ltd.

We have now developed these three challenges, and we present the opportunities to provide solutions that we might see being applied to obtain global food security in the future. Food security is not a recent demand of humanity, and the current food security issues emerged in the world press during the so-called food-price spikes that occurred between 2005 and 2008 and occupied the thoughts of many policymakers. The drivers of these spikes were many-fold and included transitions in affluence and the demand for livestock protein products along with the diversification of agricultural products into chemical and fuel sectors. The issues were those of transitions in lifestyle quality within populations, and as with many transitional movements, a new worldview is eventually reached. The new worldview has been that the opportunity of sustainable intensification of food production has emerged whereby crop and livestock production is maximised for a given area of land, and this is achieved sustainably by utilising the most efficient use of resources available to do this.[1] This world-view needs to be integrated with manufacturing, processing, retailing and consumer use of foods so that maximum efficiency of the use of resources per unit of production is realised.

This situation is very different to the worldview of the 1970s on food security where there was a rightful desire to remove hunger, and as Indira Gandhi, the then Prime Minister of India analysed it, to obtain 'a world without want'. Gandhi put forward the issue of food security backed by social–political change as one that would remove the scourge of hunger and starvation globally and provide a fair standard of living for every global citizen.[2] The essays that come from this period did drive the humanitarian need for food security into the public domain and changed our view of the food system for ever. They highlighted the role of natural events and disasters in creating famine because the centres of poverty and hunger were associated with river deltas and valleys,

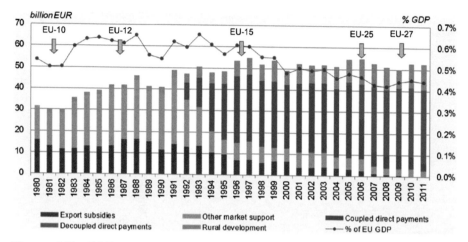

Figure 1.7. CAP expenditure and reform path budget evolution; while overall (unadjusted) budget has increased, it is worth noting the changes in distribution of the main measures (specifically the increase in Direct Aids and Rural Development), as well as the significant increase in number of farm businesses, together with the enlargement of the EU.

Source: The graph is from the DG Agriculture and Rural Development, Agricultural Policy Analysis and Perspectives Unit © European Union (2013). http://ec.europa.eu/agriculture/cap-post-2013/graphs/index_en .htm (accessed 22 April 2014).

Global Food Security and Supply, First Edition. Wayne Martindale.
© 2015 John Wiley & Sons, Ltd. Published 2015 by John Wiley & Sons, Ltd.

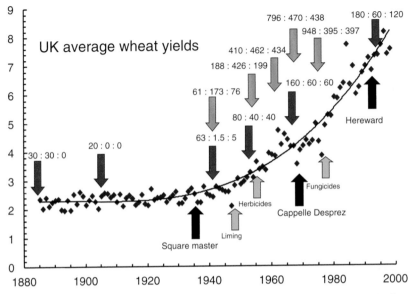

Figure 2.1. The development of agronomic innovations and their relationship to average wheat yields in the United Kingdom. Red arrows show the recommended N:P:K fertiliser application (the fertiliser recommendation for wheat in kilograms per hectare) at that date, the actual amounts of fertiliser as N:P:K applied is shown with blue arrows (in thousand tonnes). Development of different wheat varieties are shown with black arrows, and new agronomic techniques are shown with green arrows. The graph demonstrates the agronomic development of the wheat crops and the innovations that have made dramatic grain yield increases possible.

Source: The wheat yield data were obtained from FAO. FAOSTAT (2012). Production, crops (data set). http://faostat.fao.org/ (accessed 22 April 2014). Fertiliser data were adapted from the data presented by Cooke (1977),[46] and agronomic practice data were developed by W. Martindale as part of an OECD Cooperative Research Fellowship (2001).[47]

where changes to seasonal rain and irrigation resulted in flood and drought events that proved disastrous for maintaining food security by agricultural production. It is useful to consider this viewpoint because the natural events and disasters still occur 40 years later, but international cooperation and distribution technologies have ameliorated the impacts of these events and the food security threats in the twenty-first century are different. Climate change now dominates the projected changes in land use, and the impact is far different in that the effort to tackle climate and environment change has become coordinated internationally because the impacts will affect all nations.

The impact of hunger and the scale of natural disasters have not decreased since the essay of Gandhi in 1975, but the access to technology and information has transformed how we are able to tackle natural disasters whether they are prolonged drought, delayed monsoon, or flood. The definition of want has always been recognised as variable, and we can assess food security as a balance of nutrients that will provide health and well-being at the most basic level of sustaining a healthy life. However, this scenario is made far more complex because of the social and cultural influences on food consumption and the preferences of taste in populations. In her 1975 essay, Gandhi defined three wants that drive demand, and they are the following:

[To cover the] 'shortage of the essentials' for existence such as minimum nutrition, clothing and housing
[To develop the existence of education and recreation that] 'give meaning and purpose to life'.
[To obtain extra resources that] 'advertising proclaims as necessary' to a good living.

The approaches to tackling food security in the 1970s opened up worldviews on fairness and justice, with assessments of land availability and other resource availability being discussed internationally. The developed world had more of everything, notably more energy consumption and available land use; this type of assessment was the infancy of the environmental footprinting movement, which developed a range of illustrative methods to describe how much resources is embodied in a particular product or citizen. This approach has clearly shown inequalities in the distribution of global resources, but we have begun to solve where we should start to respond to reducing inequality. The past is enlightening and it can be said progress has been made by integrating new technologies and information sources as they have become available.

Thus, the development of a global food security strategy has evolved from many ideas and theories. The worldview described by Gandhi was one of aspiring to want and possibly want more than was necessary. We might consider this worldview proof that the Malthusian projections were right and population did indeed outstrip the potential for natural resources to supply enough for survival. The worldview of Reverend Thomas Malthus was established at the beginning of the industrial revolution in the United Kingdom, when the first industrial cities were developing and population transitions were evident with regard to lifestyles and wealth creation. The lifestyle transitions were written about in a similar way that they are now and were then being observed and communicated by journalists and writers such as Charles Dickens.[3] This was an age where Newtonian laws of science were being applied to industry and economics. Notably, the French financial system was being influenced by Jacques Turgot, who reported progress in terms of straight line predictions to illustrate growth and wealth creation.[4] This was a time of science and projections that provided solutions for supplying growing populations with access to perceived unlimited resources required for wealth and health; limits were neither conceivable or reachable in the world-view of that time.

4.2   Free Trade Transitions into Sustainability

The eighteenth and nineteenth centuries were a time of industrial development, the assessment of the application of science and it is no wonder the socially aware such as the Reverend Thomas Malthus saw chinks in this straight line projection worldview of progress. The Malthusian worldview was one that dominated global food policy in the 1970s with the publication of Dr Paul Ehrlich's *Population Time Bomb*. The limits to growth were well appreciated and extremely well communicated by a growing environmental movement that was obtaining ground in government and international policy. Notably the work of the Norwegian Prime Minister Dr Gro Harlem Brundtland changed our views of growth and natural resource use for ever. The very issue of sustainable development had now entered our thinking when considering the use of natural resources globally. The fragility of the global food system was predictable and characterised by experts and scientists and the problems have been well defined so that we have an opportunity to provide solutions. The emergence of defining different types of wealth creation and economic growth that went beyond the economic wealth and the straight line projections of Turgot gained pace with the global interpretations of sustainable development and wealth put forward by the environmental movement. Paul Hawken and the Lovins's aired the idea of natural capital using a sustainability tenet that you cannot continue manufacturing consumer products such as food if your resource base is being depleted.[5] There will come a time when either you cannot make more because there are no ingredients left or you have to make less because you need to make what is left last longer. Lovins also raised the issue of producing differently, and this thinking coincided with John Elkington's views on sustainability, that is, there are economic capital, social capital, and natural capital components associated with sustainable wealth creation. Even national treasuries and central banks were using this language of sustainability where the limits to growth were incorporated into economic models and risk assessment; it was no longer a case of growth being solely focused on economic capital. This established a new view of accounting where companies were not only assessed on economic profit or loss reported quarterly or annually, but also for the social and natural capital held by a company.

Indeed, much business practice is still set on short-term investments that provide fast return to investors and shareholders; this is often in conflict with sustainable approaches where social and natural capital is built into the growth of economic capital. What is more, is it is difficult to measure social capital and place this into a sustainability index that can represent the whole triple bottom line. Thus, while it seems intuitive to measure performance in terms of sustainably, it is difficult to communicate and proven impossible to measure for many organisations. No wonder the taking up of sustainability thinking took a while; the standardisation of measuring environmental performance has been stimulated by the development of international standards for product footprints and environmental management systems. This revolution has been driven by the food industry and the food supply chain, it has not been a slow starter because of prior experience with food assurance and traceability schemes.

4.3   Increasing Food Supplies Have Been a Major Achievement since 1975, but There Is Increased Resource Nationalism Evident by the Emergence of 'National Interests in a Shrinking World'

Development of modern agricultural programmes have transformed production since 1975, and the impact of supply chain is beginning to be realised, with food traceability and assurance scheme emergence. Growth strategies and resource constraint are beginning to be understood because the Green Revolution was 'a mixed picture' where technologies were deployed into markets to increase production and economic livelihoods. However, we now know that the environmental and social consequences of changing food supply are as important as applying new technologies that change production. In his book, *Earth Odyssey*, published in 1999, Mark Hertsgaard details his round-the-world journey started in 1991, providing a reflection on industrial and agricultural development with a foresight view of new developments that include environmental issues so that we might consider what the future holds for us.[6] A stark reminder of the want of the human condition at basic levels is the narrative Hertsgaard provides in his book for Sudan and the areas of the Darfur region, which was torn apart by civil war in the late 1980s. The plea of the chapter on the Darfur region, 'we are still here', describes the subsequent use of natural resources following political breakdown until all that exists is famine and an almost total loss of hope. The organisations that still exist in this situation Hertsgaard describes are the Red Cross, the UN Food Programme, and the Catholic Church and the Sudanese citizens in the narrative state despondently that 'we are still here'. This provides a stark reminder to nations who can contribute aid and assistance to do so even though technolgies and distribution infrastructure exist; there must be the political will, trust, and leadership to alleviate such suffering.

Strong leadership and trust have made significant impact on the future management of our planets natural resources; we have discussed this is terms of the people who have shaped our current understanding of sustainability. The evidence obtained from previous production and consumption trends make it clear that food businesses will largely determine what 9 billion consumers of the 2050 world will buy, use, and waste. What is not clear is how 9 billion consumers will feel about their lives or the products they consume. That is, the qualitative components of the food supply chain are becoming less well charactised than the sheer scale of LCA, energy balance, and EMS data now available. Much of these data are peer reviewed, that is, reviewed by independent panels of experts; it is also often freely available as open-access publication. Qualitative data regarding product use are often commercially sensitive and private data are owned by the producer, manufacturing, processing, and retailing components of the supply chain. Indeed, retailing has become extremely reliant of consumer data that are obtained from sensory panels, enabling products to be benchmarked against others so that product claims can be made and price points set.

These two worldviews of the corporate and consumer are documented by commentators and have become an important consideration in obtaining sustainable options in the use of planetary renewable and non-renewable resources. Humankind must find new ways of determining how appropriate corporate and consumer power in supply chains is used for different types of products. The solutions are often presented in broad assessments and scorecards of sustainability that are too complex or are not appropriate for business and consumer application. The requirement is clear to us: there is a need for an assessment of sustainability for business and consumers; if this is not achieved, it is unlikely that the sustainable goals policy-makers set will be realised. The outcome of unsuccessful assessments of the sustainability for both product supply and consumption is that policies will fail because of a lack of will to engage both business and consumer requirements together.

A consequence of delivering the values associated with culture and society we have come to expect is the removal of our lifestyles from nature, the production of foods and even the preparation of meals. There is now an expectation that at least 6 billion of the 9 billion people in our 2050 world will make the transition from rural to urban lifestyles that are associated with different cultural and social needs that often result in the distancing of people from food production. Businesses need to know what will keep the human conditions of communities communicating with each other and taking part in cohesive social functions. This needs to be achieved by communities while still remaining tuned into the food systems they utilise, otherwise food and other services derived from nature become a distant and abstract theme in their lives. A typical illustration of what we are becoming is not necessarily one of good human condition, because it may well be an obese individual, burdened with urban life, who is not necessarily happier than their ancestors were, or an individual who simply does not consume enough to live healthily. Indeed, it can be considered that we are manifestly more miserable than our ancestors, and economists will often cite the Kuznets curve, which presents the 'it will get worse before it gets better' scenario. This is where basic needs for survival evolve from a minimum requirement into the more aesthetic needs of individuals through wealth creation.

Understanding future customers, where they are, and how they utilise products is an important goal of today's foods business if we are to achieve the elusive goal of sustainability. It is clear that while labelling and accreditation systems are essential to individual product specification, they will not necessarily enlighten or change consumption behaviour to significantly change how we will utilise global natural resources. What is clear is the need to understand supply chain functions, and this can be achieved working from first principals. Knowing what our resource base is, where food products come from and how they are made is a critical requirement for sustainable consumption. Several research programmes have mapped natural resources globally that are our basis for supplying a future for 9 billion people; the volumes of these resources required to maintain future lifestyles have so far been rarely limited by localised conditions because of efficient transport and distribution systems. This is specifically the case with regard to land use and agricultural production; researchers have further quantified resources required to maintain current levels of consumption for the 2050 world where there will be 9 billion consumers.

Ultimately, the decisions inciting these transformations will depend on leadership credentials of CEOs and chief officers in companies. Since the 1960s, movements regarding environment and sustainability have highlighted issues concerning resource limitation and augmented public debate. However, a more recent and a likely to be far more powerful means of augmenting action has been a more open society with regard to the access to information consumers have on production and consumption of products associated with specific lifestyles. Indeed, we can begin to question whether the consumer will have a more dominant role in determining sustainability in the 2050 world, and there are areas of the food system where consumer trends will drive changes in supply chains. The world protein budget is critical to the future development of the food supply, and the location of where that protein comes from is critical. Indeed, it is the indicator of dietary transitions in the world, as already discussed, there are significant protein and calorific gaps in diet with regard to quantity consumed. The quality of diet is often defined by protein content, and we have already identified how the evolution of human taste preferences and protein metabolism has determined how we choose a portion of food or a balanced meal.

The data presented in Table 4.1 demonstrate the challenge that faces the global food system because we obtain protein and calories from a wide range of sources in different locations. These sources of biomass are delivered in the supply chain as ingredients that are manufactured into

Table 4.1. Global protein supply (FAOSTAT 2009 data): the components of global protein supply in terms of absolute production from agricultural primary products and their supply to individuals

	World production (million tonnes)	World protein g/cap/day	World protein g/cap/day (% of total)
Cereals	976.7	32.0	40.4
Fruits	485.5	1.1	1.4
Oilcrops	48.1	2.7	3.4
Starchy roots	406.9	2.2	2.8
Treenuts	14.2	0.4	0.5
Vegetables	877.5	4.5	5.7
Eggs	59.3	2.7	3.4
Fish, seafood	122.9	5.1	6.4
Meat	278.9	14.1	17.8
Milk	580.9	8.0	10.1

Notes: The total average protein supply for a global citizen from these components is 79.3 g/cap/day.
Source: These data were adapted from FAO. FAOSTAT (2009). Food supply, crops primary, equivalent and livestock and fish primary equivalent data set. http://faostat.fao.org/ (accessed 24 April 2014).

foods; foods are created by manufacturers or consumers as recipes and portions of meals. We will now use case studies to show how understanding the supply chain can integrate with consideration of agricultural operations, and this is essential to developing robust food security by improving nutrition and reducing food waste. There are many examples of the development of new food ingredients that do not fully consider how consumers use them or experience them. For example, we have seen that the development of golden rice was established to deal with an extremely important health impact, and sensory tests on the actual product were not carried out until development of the variety was advanced. Thus, the product was not characterised as a consumer product by consumers for several years. Indeed, this is an outcome of developing advanced novel technologies, such as genetically engineered crops; the invention is often the least complex and expensive part of the development cycle, and there is a requirement for consumer risk assessment and toxicology assessment to be carried out. These types of development cycle will change, and we will see sensory trials of products or similar products being used to develop new foods from the beginning of the innovation or research cycle.[7] Golden rice is not alone, and we must consider that most crop varieties have been developed with the focus of technologies solving a specific problem without structured research into how consumers may experience the final product. The focus of development has usually been to improved biomass yield or manufacturing efficiency.

Before we consider the ultimate part of the food supply chain, that is, where consumption of finished product occurs by the consumer, we must consider how we can support the measurement of sustainability across supply chains and the food system. The use of life cycle assessment (LCA) is central to these discussions because LCA formalises the measurement of inputs and outputs for each food supply chain function and provides the outputs in what can be described as sustainability criteria.[8–11] LCA methods present opportunities because it is subjected to increasing internationalisation of assessment standards that will have significant impact on assuring product declarations and food safety.[12] LCA methodology has therefore helped to revolutionise how food processors and manufacturers can investigate their supply chains and report their performance.[13] LCA has evolved to become more suited to assessing the sustainability criteria of fast-moving consumer goods (FMCG), and this has resulted in the emergence of footprinting standards and the use of LCA in modular forms or simplified versions to investigate specific criteria, such as greenhouse gas (GHG) emissions in the carbon footprint methodologies.[14] This approach is important because it is ultimately driven by the supply chain delivering what are likely to be more successful products. How this is assessed can vary depending on the modular form of LCA used or attributes assessed by footprinting, but data regarding consumer use will be critical in all assessments. LCA is often confused with the assessment of consumer behaviour and the use of FMCG, but it has nothing to do with consumer behaviours because the supply chain will respond to how products are consumed.

Currently, many LCA approaches overlook the value of understanding the role of consumer use of FMCGs in the continued development of new products. Manufacturers and processors are likely to benefit most from understanding these types of data because in supply chain terms, they sit between agricultural producers wanting the highest price for ingredients and retailers who want the lowest prices for customers in supply chains. Thus, they operate in a specific place in the food supply chain where the cost of resources such as energy, water, and materials is of critical importance to a business. This has resulted in the development of programmes of assessment, such as the carbon footprint international standard PAS 2050 (Publically Accessible Scheme 2050).[15] The resulting guidelines on how to enable the use of LCA thinking in supply chains and companies has been stimulated by the development of 'roadmaps' by specific industrial sectors, such as the dairy-and livestock-producing arenas.[16,17] These approaches enable practitioners in the food supply chain to begin to use LCA in a modular form and work to implement full LCA programmes as commercial demand develops. The LCA arena has become more user-friendly and commercially applicable by the use of footprinting methods for FMCGs and the need for manufacturers to understand how consumers utilise products.

Examples of LCA approaches for FMCG supply chains can be demonstrated for several food product categories if data concerned with supply chain functions are available.

Energy balance is a methodology that is an LCA approach, and it assesses the energy inputs and outputs of a system in terms of the efficiency at which energy is consumed to produce a product. It has provided important insights into manufacturing and processing operations, and it provides a very useful starting point for contemplating the complexities of the LCA, water footprint and carbon footprint of food products. Energy balance methods have been used very successfully to assess the efficiency of industrial nitrogen fixation, which can be viewed as the driver of much biomass production.[18] Whereas product carbon footprints are guiding current trends in sustainability and are likely to lead to a further wave of innovations in the food supply chain, many organisations have difficulty in how to utilise sustainability criteria of products with respect to their customers. Energy balance cuts through this indecisiveness by immediately showing where energy costs can be reduced in supply chain functions.

4.4 A Demonstration of Energy Balance and LCA for Sugar Production in Europe

In many food supply chains, the agricultural production of ingredients is a key area of GHG reduction activity because of the use of nitrogen and fossil fuels that represent major sources of energy inputs that are associated with increasing GHG emissions. Farm management strategies that deliver environmentally sustainable outcomes are well known to growers because of the impact of the set-aside and stewardship schemes established in the late 1980s. How GHG emissions can be further reduced will provide farmers and growers with future challenges in maintaining production efficiency and profitability, and these must focus on the use of fossil fuels and nitrogenous fertilisers. The sugar supply chain in Europe offers an important case study because sugar is produced from sugar beet, and in the United Kingdom, there is only one processor of sugar beet into white sugar, British Sugar plc.[19] This means the supply chain is well understood from farm production to retailing, and it provides an important demonstration of the value of energy balance.[20]

Farm practices that conserve GHG emissions are not as high profile or visible as manufacturing operations because they cannot be metered directly and are more complex in that GHG emissions are extremely dependent on annual variations in weather.[21] Even with these extremely variable attributes, we can use LCA approaches as energy balances to determine the sustainability of inputs, such as nitrogen fertiliser. Furthermore, established annual yield reports and weather trends allow us to project and standardise our analyses. With this in mind, we can begin to highlight fit-for-purpose actions farmers and growers can take in the sugar beet and other agricultural ingredient production supply chains to reduce GHG emissions for each growing season. The LCA and energy balance are methods developed by engineering companies in the 1970s, that have developed to become an international standards and the basis for undertaking a carbon footprint of products. The approaches take a whole system view of the supply chain, and it is grounded in the common sense that we do not get something for nothing. In this case, converting biomass into white sugar and co-products means there are energy transfers, and when that energy is transferred, it either becomes unusable or lower quality energy, such as steam or 'low-grade' heat within the production system. LCA and energy balance approaches identify where energy transfers occur and where energy might be recycled through the system to make the best use of it. This seems straightforward, but soon becomes complex because supply chains have many suppliers, different processes, and varying traceability.

The carbon footprint represents the embodied GHG emissions of a product across its lifecycle—from the production to the disposal phase ('cradle to grave')—and offers a convenient method for assessing GHG impacts across food supply chains. LCA methodology is used to calculate the product carbon footprint and a functional unit of the system or supply chain being analysed.[10] The functional unit in the LCA and footprint of food products is typically a specified mass of product that is used by the consumer, such as a gram of white sugar. The carbon footprint and part of an LCA determines the total amount of GHGs associated with a given functional unit. In this case, the GHG emission reduction opportunities for the sugar beet supply chain before the factory may not be fully considered on the farm even though growers are expert in optimising the sugar beet production system. A convenient way of investigating the GHG emissions associated with sugar beet production is to understand the energy consumed during a typical growing season. Figure 4.1 shows the energy balance for sugar beet production obtained from 32 years of field trial data for eight independent cereal, potato, and sugar beet rotations on different farms in Germany.[18] This German long-term study offers some important insights into sugar beet energy management on farms in that direct use of diesel by the grower makes up nearly 70% of the energy balance, and includes ploughing (22%), harrowing (7%), and harvesting (40%).[22]

The energy balance shows us that the total energy input for sugar beet and other root crops such as potatoes is typically 25 GJ (giga-joules) per hectare and typical energy outputs of the sugar beet crop are 350 GJ per hectare with co-products included in this figure with white sugar yields. The energy output for only white sugar was typically 250 GJ. This represents a 10- to 14-fold increase in energy output due to efficient agronomic management, ensuring the sugar beet crop canopy captures as much solar energy as is possible during the growing season. There are important opportunities that need highlighting from the energy

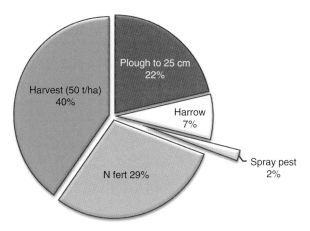

Figure 4.1. The energy balance of sugar beet production: percentage of total energy input associated with different production inputs and harvesting, using data derived from the 32-year long-term field trials of eight potato–cereal–sugar beet trials in Germany 1967–1998.
Source: **Data from Hülsbergen et al. (2001).**

balance that result in lower GHG emissions and reduced production costs. It is notable that ploughing and seed-bed preparation have changed in recent years so that diesel consumption is reduced. The direct and indirect use of energy for the operations identify that at least 50% of the energy used for ploughing, harrowing, and harvesting is under the direct control of the farm through consumption of liquid fuel. This means improved efficiency and design of machinery and fuel consumption will have a significant impact on the farms sugar beet production energy balance. An important means to reduce fuel consumption is the consideration of minimal soil cultivations that reduce the need for plough cultivations and utilise discs and harrows only to produce a seed bed.

Of course, minimal cultivations may not be appropriate for all soils and weed management regimes, but there are often options to reduce field traffic and soil cultivation intensity across a cereal-beet rotations to reduce GHG emissions and increase profit margin. The development of cultivation machinery that combine discing, harrowing and pressing cultivations into a single or reduced pass for seed-bed preparation has had impacts on GHG emission and energy balance of sugar beet production that should be regularly measured and reported. The message from the energy balance here is to reduce ploughing and field traffic as much as possible because the benefits are both financial and environmental. Whereas minimal cultivation may reduce the energy consumption of seed-bed preparation, it is unlikely that there will be a minimal cultivation option for sugar beet harvesting. However, considering harvesting as a soil cultivation in the rotation is measurable, and the fact it is responsible for 40% of the energy consumed it should be accounted for in the benefit to the following crop. Thus, there are opportunities to reduce GHG emissions of soil cultivations, and using biodiesel will reduce them further because biofuels recycle GHG emissions through the farm because the fuel is both grown and combusted on farm, making fuel consumption close to carbon neutral.[23] A further benefit of minimal soil cultivations is it will stabilise the soil organic matter content, increasing the amount of carbon fixed in the soil, which has been the thrust of soil carbon conservation programmes in the United States, where minimal cultivations have transformed combinable crop production through the market entry of herbicide-resistant crops and machinery innovations. Research reported by Rothamsted Research shows that up to 10.4 million tonnes carbon per year could be fixed by UK agricultural soils using minimal cultivations, which is equivalent to some 2% of the national GHG emission inventory. What is crucial to understand is that these energy efficiencies and GHG conservation measures can be passed on to the food supply chain and eventual consumer use of a product, such as white sugar.

A similar scenario exists with the use of nitrogenous fertilisers, whereby changes in the management of the production of biomass that result in benefits can be passed onto other food supply chain functions. The use of nitrogenous fertiliser accounts for 30–35% of the energy balance for crop production on farms because of the energy required to fix nitrogen in nature or by the industrial Haber Bosch-derived processes. The energy required to fix nitrogen to ammonia has decreased to near the theoretical minimum required, and it is therefore important that we find new routes to conserving energy in the farm system.[24] It is the fixation of industrial nitrogen that supplements natural nitrogen fixation, and Smil identifies that 24% of all nitrogen consumed is industrially fixed, so we cannot do without it, but it does come with impacts that increasingly require further optimisation of nitrogen use.[25] The energy consumed for nitrogenous fertiliser, whether as mineral (principally ammonium nitrate and urea) or animal manure, is approximately equal because we must consider the energy consumed to produce, transport and spread animal manures.

The energy, nutrient, and GHG balances will always show us that we can never get something for nothing, and in the case of crop nitrogen fertilisation, we have another factor, that is, the complex relationship between effective nitrogen fertiliser use and the yield and quality of the sugar beet crop. Increasing nitrogen fertiliser will increase biomass yield, but there will be a point where the quality of sugar beet is compromised because of the impact of increased molassogenicity of the beet. This is caused by nitrogenous compounds in the beet that increase as nitrogen fertiliser is increased, with the resulting reduction in white sugar yield during processing and manufacture.[26] Similar types of relationship between nitrogen fertilisation and crop quality are observed for the following.[27]

1. Bread wheat where increased nitrogen fertilisation improves protein content and bread making quality.

2. Potato size and skin set can be influenced by different nitrogen fertilisation and determine different marketability, for example, small or salad and large or table potatoes.

3. Malting barley where decreased nitrogen in the grain provides more suitable malt for beer making because there will be lower protein derived nitrogen in the beer produced. However, the presence of protein is important in producing foams that are associated with the consumer experience of beverages, such as beers and coffee.

As Smil (2002) explains, the energy required to manufacture mineral nitrogen fertilisers from industrially fixed ammonia are close to the minimum theoretically possible. This means that the management of factory emissions, the fertiliser supply chain, and farm use of mineral fertilisers are the only realistic options for significant GHG for energy balance or GHG emission savings to be made. The carbon footprinting of nitrogen fertiliser products has been reported with GrowHow UK Ltd Group of companies footprinting its nitrogen products with the carbon label standard (PAS 2050), and it is likely that other nitrogen manufacturers will follow.[28] The important consideration here is that of the major energy input for the use of organic manures; the use of fuel to transport bulk manures is as important as those fuels consumed in soil cultivations. That is, reducing field traffic and the transport of materials, which result in reduced diesel consumption, will reduce the carbon footprint of nitrogenous fertilisers, and in combination with these efforts, we must consider the use of biofuels for farm operations so that GHGs are recycled within the farm system. Thus, transportation of products and ingredients has a significant impact on the sustainability criteria of FMCGs because fossil fuels are currently intensively consumed, meaning we can either change transportation fuel sources or optimise transport routing.

Transportation has been identified across supply chains as an area where improvement is often possible and needs to be integrated with improved product and manufacturing processes, as well as storage, wholesaling, and retail functions of the supply chain. The use of geographic tools offer solutions because they can help us plan for transport costs strategically and assess the energy balance, GHG emissions, and costs associated with transportation and other supply chain operations.

4.5 Carbon Footprinting for Food Manufacturers Begins to Offer a Sustainability Reporting Framework

The embodied GHG emissions (the 'carbon footprint') of a product across its lifecycle—from the production to the disposal phase ('cradle to grave'), offers a convenient criterion for assessing GHG impacts and possible sustainability criteria of products across food supply chains. The method it is based on, LCA methodology, is used to calculate the product carbon footprint, and a functional unit of the system or supply chain is analysed so that the carbon footprint determines the total amount of GHGs associated with a given functional unit. PAS 2050 (2008) offers the potential for greater understanding of life cycle GHG emissions by the consumer within a context of making purchasing decisions between goods and services transparent with regard to GHG accounting. This suggests that carbon footprints of food products will be increasingly used by industry to report dietary criteria associated with health and social responsibility as well as GHG responsibility. That is, the more we eat them, the more GHG emissions result and, the more we waste, the more GHG emissions are consumed. They will not provide a total sustainability index for whole organisations or supply chains in the food system, but they will provide an important starting point in achieving the goal of accounting for sustainability criteria in supply chains.[29] Indeed, the wider goal of developing indices that integrate consumption and sustainability criteria have been tested by food retail groups and represent a future challenge to food businesses (e.g. the Walmart Sustainability Index[30]). Recent research by the author and Iglo Food Group Ltd has shown that the use of meal planning and frozen foods can reduce food waste in the home.[31] That is, appropriately preserved food enables consumption that results in less food waste. The study uses novel questionnaire methods that determine a food waste index which relates food preservation, meal planning, and amount of food waste associated with purchased foods. This study was achieved with a sample population of consumers trained for sensory panel analysis and recording food consumption attributes.

Protein supply is crucial to the future sustainability of the food system, and the major sources are currently cereals, pulses, and fish and livestock products. The role of other protein sources, such as those fermented from carbohydrate, is relatively unknown, and mycoprotein provides an important case study in reducing livestock product consumption while maintaining the protein content of meals that might be consumed in future.[13] The future global protein supply market is uncertain with respect to maintaining sufficient protein supply for a global population that is projected to reach 9 billion in 2050, and this will have important impacts on product development and international trade. This is because of the increase of the more affluent purchasing in the economies of Brazil, Russia, India, and China changing how protein is traded globally. While the impact of this affluent purchasing power has been described in terms of increased demand for livestock products, the issue of developing sufficient protein supply is clouded with many factors that are far more complex than a meat versus non-meat choice.[32] Food consumption is a significant proportion of national GHG emission inventories that are reported for policy purposes, and in the United Kingdom, for example, the food and drink system is responsible for 195 million tonnes of GHG (CO_2e) emissions per year; these are carbon dioxide equivalents and include all GHGs of current policy interest.[33] Thus, using individual carbon footprints of products to assess the GHG consumption impact of the dietary behaviour of populations will be an important consideration for environmental policy that acts throughout the supply chain.

The analysis of the impact of food consumption at national scales with respect to reducing GHG emissions is still an emerging area, and actions that are likely to result in GHG conservation across the food supply chain are often untested. One such action may be a reduction in the consumption of livestock protein; while it is uncertain how changes in the consumption of livestock products will reduce GHG emission, it is increasingly clear they have an important role to play.[34] The use of industrially synthesised or fermented proteins for human consumption offer an important option in determining what future protein supply may be; indeed, how the distribution of protein consumption, shown in Table 4.1, might look like in the future.

The development of mycoprotein is such a fermented protein, and it provides an important case study in protein sustainability here, with its story that starts with a collaborative partnership with the RHM Industries food group and Imperial Chemical Industries (ICI) in 1984 that formed Marlow Foods Ltd. This company was a mycoprotein manufacturer that developed the current 155-m^3 scale air lift fermentation technology that grows *Fusarium venenatum* under strictly defined conditions in the United Kingdom. Harvesting of mycoprotein from fermenters can continue in steady-state continuous flow for up to six weeks. At this stage of manufacture, mycoprotein has the appearance of bread dough but lacks the elasticity associated with a bread wheat gluten mass. For the production of Quorn™, the branded product of mycoprotein, the mycoprotein is mixed with egg albumen, roasted barley malt extract, and water.

Mycoprotein does offer an alternative scenario to livestock proteins in that a high-quality protein is now available by fermenting glucose into protein providing an ingredient for food products that is 11% protein (w/v) containing all essential amino acids.[35] The manufacturing and processing innovations associated with the development of mycoprotein has resulted in the Quorn, which includes shaped products, such as fillets, mince, and pieces, coated ready meal products and ready-to-eat deli style products. Quorn itself is produced by including 8–10% egg white protein to mycoprotein, bringing the protein content of Quorn to 14–16% w/w in finished products. There is significant interest in lowering the livestock product content of diet because of the health and environmental implications that have been outlined, and there is now a growing body of public advice that provides ways to decrease livestock product consumption. For example, the World Wide Fund for Nature (WWF) has developed the Livewell Diet, which proposes that small reductions in meat consumption per meal can lower the GHG impact of food consumption and promote healthier lifestyle.[36] The global perspective for protein supply provides a requirement for national agencies to assess supply chain efficiency because increased protein demand pressures are currently largely dependent on feed and livestock production systems.

The carbon footprint calculated for mycoprotein and selected Quorn products provides an important demonstration of a company utilising sustainability criteria to explain the impact and benefits of using their products in meal by consumers. This has been guided by the requirements of PAS 2050: 2008, the Code of Good Practice for Product Greenhouse Gas Emissions and Reduction Claims (2008), and the Carbon Trust's Footprint Expert™ Guide. The development of the carbon footprint for food products is becoming relatively simple for manufacturers to undertake because the recording of supply chain data is usually undertaken by manufacturers as part of standard audit practice, and it is now becoming a case of converting this information into GHG emissions using published conversion factors. Whereas this activity used to be only done by researchers, it as now expanded to being used by practitioners who regularly record data for waste inventories or to determine economic costs of manufacturing. The process of data analysis in supply chains is becoming one of establishing data collation policies and using appropriate data conversions for economic, social, and environmental interpretations.

In a manufacturing context, primary data (directly measured) and associated secondary data (derived from conversion factors) for the carbon footprint of a product are derived from direct measurements of ingredient volumes, metered electricity consumption, metered steam consumption, and metered effluent treatment volumes used during the production. For the raw material data process input and output volumes, processing energy requirements, logistics, and distribution processes, the internally collected company information can be collated to provide an important part of the environmental or CSR review within a company. Energy, steam and water use data are usually collected using metered records to specific manufacturing processes, such as aeration, steaming, chilling, and freezing in the manufacturing and packaging stages. The volumes of all inputs, outputs, and waste associated with the finished products are used to obtain the carbon footprint. Process maps guide the procedure and are placed into the context of the data collected for carbon footprint procedures.

4.6   What Can We Do with Sustainability Assessments of Food Products? Using Carbon Footprint Data in Supply Chain Management

The data sets we can now use to determine the sustainability criteria of food products has increased because LCAs for product groups are now reported widely. The use of LCA in directing dietary and health policy is also described, and this has resulted in the consideration of the carbon footprint of different diets. Databases of LCAs now exist for the production of agricultural products, and the data concerned with manufactured food products remain relatively limited but they are growing.[37] The value of considering LCA in the food processing and manufacturing arena is still developing, and it does offer many opportunities to businesses that aim to utilise resources more efficiently and need to report sustainability criteria of their products.[21]

Indeed, the reporting of sustainability criteria associated with agricultural products is likely to be more readily available because they are not strongly associated with branded products. The carbon footprint arena has become more competitive, as owners of food brands observe competitive advantage in reporting sustainability criteria associated with products. The stage for footprinting and LCA has changed from understanding the supply of food ingredients and staple agricultural goods to one of measuring to performance of brand value. The impetus for carbon footprinting or carrying out LCA for specific branded products will emerge from individual companies and organisations, but they can still utilise published LCA and GHG conversion factors for standardised ingredients. There are established LCAs reported for branded food products, such as the mycoprotein case study described here, and other business supply chains, such as that for white sugar.

The carbon footprinting of food products has also raised the issues associated with water use in the food supply chain. This is most important in manufacturing and processing sectors because the generation of steam and utilisation of heated water to clean, sterilise, and form foods is critical. The uses of water in the food supply chain are often realised by undertaking a footprinting or LCA programme within a company. The scope to develop LCA methodologies to encompass water use with GHG emission criteria will provide opportunities for food businesses to identify and report resource use efficiency in the future.[38,39] While water use will provide many challenges to the food industry, the stimulus to develop and conserve water use in a similar way to energy use in the manufacturing sector is gaining momentum.

The use of carbon footprinting is particularly useful in considering the value of protein in food supply chains. Crops provide an important protein component of meals and animal feeds; the principle crops used as protein sources are shown in Table 4.2. These data were obtained from published LCA data and the FAO statistical database, and it can be seen that crops have lower global warming potentials (GWPs) associated with their production compared with livestock products shown in Table 4.3. The

Table 4.2. The GWP, land use, and production attributes of crops

Protein source	GWP kg CO_2-eq/kg	Land use ha/t	Production (M tonnes, 2010)
Wheat	0.5–0.8	0.5	653.7
Soy bean	0.6	0.3	265.0
Rice (paddy)	2.8	0.5–0.8	696.3
Maize	0.5	0.5	840.3

Source: The GWP and land use data were obtained from published LCA databases,[40] the production data were obtained from FAOSTAT. These data were adapted from FAO. FAOSTAT (2012) Production, crops (data set). http://faostat.fao.org/ (accessed 24 April 2014).

Table 4.3. The GWP, land use, and production attributes of livestock

Protein Source	GWP kg CO_2-eq/kg	Land use ha/t	Production (M tonnes, 2010)
Beef	4.0–38.4	4.0–9.0	64.1
Pork	4.6–6.4	0.6–1.2	109.2
Chicken meat	3.1–4.6	0.50	86.5
Eggs (chicken)	1.3–1.9	0.4	63.8
Milk (cows)	0.9–1.8	0.07–0.09	600.8

Sources: The GWP and land use data were obtained from published LCA databases,[40,41] egg production data were obtained from the Australian Egg Corporation Limited,[42] and production data were obtained from FAOSTAT. These data were adapted from FAO. FAOSTAT (2009). Food supply, livestock, and fish primary equivalent data set. http://faostat.fao.org/ (accessed 24 April 2014).

GWP of a product is equivalent to the carbon footprint, and it represents the GHG emissions associated with the production of a specific mass of a product.

4.7 The Interactions between Affordability, Accessibility, and Food Security

We now understand how the environmental impact of protein consumption can be changed, and we need to also comprehend what are the structural and market factors that affect individuals and households in terms of access to, and affordability of, a healthy balanced diet, and what policies and interventions are effective in managing these. Dietary choice is a key attribute to understand and measure here because it will determine the outcomes of consumption and waste.[31] Obtaining a suitable way of measuring food choice is a difficult task to achieve because ideally the method needs to be utilised internationally across all food supply chains. The structure of supply chains change dramatically globally, and the arena in which consumers purchase and choose foods will be variable. An understanding of this variability is important, and standardising this for analysis must be made possible; the density of purchases will be critical to any analysis. The analysis of purchasing is made possible by developing maps of the retail landscape to develop the format in which they are presented so that they are fit for purpose in production, manufacturing or retailing. The data regarding volume and density of goods sold by the retail outlets and the food categories sold as fresh produce, frozen foods, processed foods, ingredients and ready meals will be an important data source. However, how consumers use food products in the domestic environment is just as critical to reporting sustainability criteria. In terms of affordability and accessibility, these are made possible by the distribution and retailing functions within the supply chain, and they are of most interest here. The efficiency of food distribution and preservation are crucial to delivering safe food to consumers, and this can be demonstrated using case studies from efficient food supply chains. Preservation of foods and the development of cool chains for refrigerated and frozen goods are currently essential to achieve accessibility and affordability. The maintenance of cool chain activity is energy intensive, and it has significant sustainability impact because it is an important source of GHG emissions, and it also reduces the waste of food products by preservation of efficient distribution, retailing, and domestic use of products are achieved.[13]

Food waste trends have identified that most waste is produced by domestic use of food in developed supply chains and prior to processing or manufacturing in undeveloped supply chains. The resources utilised in delivering cool chain FMCGs is a significant target for future actions in developed supply chains because they are critical to reducing food waste by the consumer. Using the Carbon Trust refrigeration road map data, an energy balance of the frozen food supply chain has identified transport and retail sectors account for 64% of GHG emissions associated with the cool chain prior to domestic use of fresh or frozen foods.[13] Fit-for-purpose food preservation facilitates the reduction of food waste produced by consumers in preparing food, and even though the cool chain is responsible for up to a third of the energy used supplying food products, preservation methods such as chilling and freezing do conserve food waste.[31] Food waste minimisation across food supply chains is identified by many policymakers as an area where many opportunities for reducing environmental impacts exist.

The implementation of waste minimisation using 'no or low' technology interventions are proven across industry, and formalised waste management planning has become a standard protocol in many food companies. Provision of efficient mechanisms for waste management technologies should be appropriate for how consumers use products. The impetus for embedding waste management in food processing and manufacturing environments has been encouraged by the economic benefits of reducing waste. The food processing and manufacturing industry has specific challenges because large volumes of waste are organic and biodegradable. The establishment of anaerobic digestion, pyrolysis, and incineration technologies associated with solid food waste supply are now commercially proven. There is also a business requirement to divert 'waste' streams to valuable co-products, and these types of actions generate new ideas, increased wealth and continued regulatory compliance.

Co-product markets are now proven with many of the supporting technologies coming from the ingredients industry. They include starch by-products, biofuels, fibres, novel oils, waxes, cellulosics, and a range of fine chemicals from biorefinery systems. Significant studies identify that at least 20% of food purchases are not consumed and they are disposed of, and that reducing consumer food waste will have important implications for the processing, manufacturing, and retail sectors with regard to product pricing and design. The food industry will need to accommodate significant challenges regarding waste reduction, because the pricing, marketing, and design and portion control of products will be influenced by it, and that necessitates a whole food system approach. An opportunity for reducing consumer food waste is to consider the role of preservation techniques in the supply chain with longer-term preservation methods, such as freezing, high pressure treatment, and irradiation. Longer-term methods of preservation will need to ensure nutritional, taste, and organoleptic properties of foods are not compromised, and this is a focus of future innovative food research.

We have already considered the requirement of the food supply chain for water, but this will have important impacts of accessibility to food because water scarcity is greatest where the requirement for water availability (or water stress) is most intense globally. For example, temperate climates are likely to experience a more Mediterranean or semi-tropical climate in the future representing a challenge for food manufacturers. Researchers are developing methods that measure water used by food products, and it is relevant to describe the principle of a water footprint because it develops LCA to take a spatial approach because of the variability in the distribution of water resource stressors. There is a requirement to understand how water resources are used in food systems and what methodologies are fit-for-purpose to measure the water footprint of processes associated with foods. The water footprint measurement is achieved by identifying rechargeable and non-rechargeable water resources that are associated with food products. The method of calculating the water footprint as solely embodied water regardless of where it comes from is flawed, because it only refers to the total volume of the water used in the product life cycle and does not take into account the type of water used, for example, 'green' (rain water) or 'blue' (water from rivers and reservoirs). Furthermore, embodied water footprints do not consider if the water comes from water-stressed or water-sufficient areas, where the impact of water use in an area where there is an abundance of water is very different to a water-stressed area.

Ridoutt and Pfister have provided a revised method of calculating the water footprint of a product by taking into account the Water Stress Indicator of the area where the water is used.[38] This correction gives a much improved result for the environmental impact of making that product and represents a system wide approach that depends on geospatial data. A future challenge for the water footprint methodologies is the use of spatial information that quantifies and identifies where different types of water resource exist. The implications for the management of water resources are complex and difficult to resolve because private companies, public organisations, government organisations, and consumers have different commercial and social investments associated with it. Spatial information for water resources is likely to become more valued and strategic, it also relates to developing consumption maps of the food supply chain. Research has already demonstrated that the geo-spatial information associated with food brands, food manufacturing processes, and production of waste can radically change the way food manufacturers use water. That is, the manufacture of products that use water intensively is located in areas of low water stress to alleviate regulatory pressures on manufacturing and provide improved sustainability criteria associated with products.

The methodologies used to deliver sustainable food manufacturing and processing can focus on specific aspects of product design, such as the established accreditations for carbon and water. However, there are emerging requirements to map the impacts of products across supply chains and populations of consumers. This approach enables a more detailed investigation of consumption to be completed for the life cycle of a product because maps can combine several sustainability criteria to produce indices of consumption impact and dietary choice. The development of robust evidence databases for resource efficiency has enabled a greater understanding of products and their consumption. This is particularly important when accounting for the environmental impacts of diet, which represents a specific challenge to the food processing industry because the footprinting of individual products may not correlate to the footprint of whole diets. Understanding how individual products are used and consumed will provide significant opportunities for the processing and manufacturing parts of food supply chains.

We have seen that the management of world food supply is clearly not a yield issue alone because nutritional value, consumer trends, and the infrastructure of the supply chain in the food system are of key importance. The future food system must investigate the influence of managing intrinsic supply dynamics of food and beverage ingredients and products in order to manage consumption for the desired outcome of increased global health and well-being. The global food system has been challenged with the sustainability challenges of reducing greenhouse gas emissions, reducing food waste, and improving health and well-being attributes of products, and we have shown that manufacturers can provide solutions to protein supply and waste reduction. The food processing and manufacturing sector is critical in delivering solutions to these challenges that are increasingly enforced by policies and regulations that aim to meet national greenhouse emission targets. Understanding how the food supply chain operates between the agricultural and processing segments is critical to providing secure supplies of foods because significant preservation opportunities exist here that will deliver healthier foods and reduce food waste in the supply chain. In order to develop a sustainable food system, it is necessary to use whole supply chain approaches to measure the impact of producing and consuming food products. An important aspect of understanding how supply chains perform is communication of data regarding inputs and outputs throughout the supply chain. These input–output variables can be directly measurable as costs and volumes of resource, or indirectly measured, as in the case of trust and leadership, which have not considered here yet. The measurement of supply chain efficiency will depend on the acquisition of numerical data regarding resource use, but there is also a need to consider how organisations will use these data to provide leadership.

The global scientific community has excelled at creating an inventory of global renewable and non-renewable resources that was borne out of a necessity to account for primarily fossil fuel and food production reserves. There is a requirement to reduce risk in utilising these inventories by businesses and the development of methods, such as footprinting and LCA, that try to understand demand and consumption in economic, social, and environmental terms has transformed the potential application of inventory-based models for application in policy, sustainability, and business. The data collection and processing capability required to develop robust models of product consumption are set to transform how businesses operate. The opportunity to develop business for both profitable and sustainable outcomes is now attainable in a world where it is currently well characterised that transformations in lifestyle are largely delivered by businesses and corporations, not consumers. Ultimately, the decisions inciting these transformations will depend on leadership credentials of CEOs and the chief officers of companies, that is, through appropriate leadership that is often distributed through an organisation. Since the 1960s, environmental movements have regarded many of the issues we now group into the sustainability agenda as a goal of businesses; in doing so, NGOs have highlighted these issues and augmented public debate. However, a more recent and likely to be far more powerful means of augmenting action has been a more open society that has improved access to information on production and consumption of products associated with specific lifestyles. Indeed, we can begin to question whether the consumer will have a more dominant role in determining food sustainability in the 2050 world. These issues are the concern of the retailer and the consumer part of the supply chain where retailers aim to present ideal choice to consumers, and they respond to trends in health, sustainability, quality, and value.

4.8   Retail, Distribution, and Wholesale

The development of efficient food supply chains for populations who want products that meet value and quality attributes required is fraught with difficulties and presents us with a taut balance of safety, environmental impact, quality, and choice that results in a need to know where the risk of supply failing the consumer must be calculated. This statement might seem of some relevance considering recent adulteration of beef products with horse meat DNA in the United Kingdom in early 2013, which exposed weaknesses in stating 100% assurance in light of increasingly complex supply chains and improved analytical standards. Traceability associated with food safety has been transformed over the last 20 years, and we do have an accurate view of what the distribution network of food products globally looks like.[43] There are areas of uncertainty associated with current assessments of global food distribution that need to be improved, but they do provide analysis of critical points in distribution, where compromising food safety and assurance will result in greatest risk.

The risk of adulteration and contamination in food products in efficient supply has been apparent since the rise of urbanisation and concentrated manufacturing. For example, legislation that protects the consumer of foods was forced by crises throughout the nineteenth and twentieth century in the United Kingdom. The resulting crises themselves were acute and shocking, such as contamination of sweets with arsenic in 1858, killing 20 people; more recent is the contamination of dairy products in China with dangerous melamine, which increased perceived protein content of foodstuffs. The age of health and safety reform through science and a populous link to improve the plight of the commons was a goal of much of the mid-19th century with the resulting establishment of public health policies. Outside of the acute contamination of food crises, the advances in public health were often driven by crises associated with water quality. Notably, Dr John Snow found the link between water quality and cholera in 1850s London, UK using maps and established initial epidemiological studies in crisis management.[44] These provided a foundation for our current studies of traceability in supply chains, which are in principle similar to epidemiological scenarios where the result is consumption of foods that have varying consumer responses in terms of perceptions of quality and taste.

A common link in all of these food crises of adulteration or contamination are the drivers of price and availability, whether that is the producers and manufacturers wishing to drive down prices of ingredients or consumers unwilling to pay increased prices for food products. Where and at what, intensity the price pressure is exerted will often determine the level of adulteration risk where there is high demand for products. Of course, opportunity of obtaining access to the supply of cheaper ingredients will change the marketplace, and the opportunity to adulterate is dependant on the establishment of trust within the supply chain. This presents a very clear picture of where we are now because traceability processes are visible to all of us in the food supply chain. Issues of trust, justice, and responsibility are far more visible now than they ever were even if we do not necessarily recognise them on individual products as consumers. These processes of assurance and traceability have transformed the food industry, but risks still occur, as is evident with horse DNA contamination. Of course, the opportunity to criminalise the supply chain will always be a threat, but trust and brokering trust along a supply chain was once thought impossible in global supply. We have proved it is not if we just consider the mechanisms by which we can trace food from farm to fork.

The use of labelling schemes, such as those already described, have now developed to include sustainability goals, such as the GHG emissions associated with the consumption of products. The issues of fairness and trust have been aired in food supply chains with respect to animal welfare attributes, which have proved decisive in the sales of products such as free-range eggs and those products that include them as ingredients. Egg protein is a good example of an ingredient that exposes the impact of uncertainty and risk in food supply chains, as it is used in a huge number of foods and animal welfare issues have transformed production and sourcing policies globally. Thus, traceability is evident, but crises still exist because of uncertainty and variation in trust. The consumer does not necessarily feel protected by the law in such an uncertain environment, since many Europeans state that the food that they eat is unsafe. This is a ludicrous situation given the development of our food supply chains. A solution is twofold in that greater understanding of food supply by consumers and greater surveillance of risk in supply chains is required. Each retailer uses different criteria and ultimately depends on responsible reporting of qualitative data focussed around consumer choices.

Understanding the critical role of distribution networks in developing sustainable food and beverage supply chains has resulted in many case studies of what sustainable food distribution can do. There is a dearth of information for emerging economies where the establishment refrigeration and cool chains are still developing and in many respects is catching up with the successes of efficient production provided by transitions such as the Green Revolution. This is currently most obviously manifested as food waste arising before food manufacture or processing stages of the food supply chain, and this offers many challenges for the future sustainability of the food system.

A large study reporting how LCA and Geographic Information System (GIS) methods can be used to assess the sustainability of food transport for a region has been established by the author in the United Kingdom, and this is used as a case study for sustainable food distribution here. While LCA and GIS can provide scenarios and demonstrations of food transportation, a critical part of any distribution function is identifying and delivering innovations that make food distribution more efficient in terms of price, product quality, and customer experience. The case study described here is for the Yorkshire and Humber Region in the United Kingdom, and it delivered sustainable distribution options for a group of 60 small to medium companies (small and medium enterprises, SMEs). The total sample for the Flow Project included eight detailed case studies and the results for 52 companies presented here.

Understanding distribution patterns for food and drink supply chains is an essential prerequisite for implementing logistical frameworks that aim to provide sustainable distribution systems that enable efficient business development. Sustainable food and beverage distribution can implement many innovations because of the relationship to timeliness, packaging, fuel consumption, and assurance among many other attributes. For example, these may include designing out waste to conserve fuel and space; development of novel preservation and packaging to extend shelf-life; utilisation of biofuels globally and transport efficiency tools; and implementing customer relationship management (CRM) frameworks that stimulate cooperation between suppliers in a distribution network. Table 4.4 shows the initial research in this case study, which determined the current state of food distribution operations for 52 SMEs and shows how innovations might be implemented to improve food logistics. This type of initial survey provides a snapshot of supply chains, and it can identify where implementation of relevant and appropriate technical innovations across whole supply chains can be deployed.

The limits of regional agricultural product supply have been traditionally removed by efficient food logistical infrastructure, preservation, and packaging. Globalisation of the global food system has entailed the development of novel methods that assess the efficiency and security of supply chain function. The requirement to understand efficiency has been coupled with the requirement of policy-makers to define sustainability and health criteria associated with food supply chains. We are now beginning to account for these limits using LCA and footprinting methods, which have provided a need for food companies to identify fit-for-purpose data sources and data analysis methodologies that deliver sustainable assessment of transport. What has not been tested rigorously is the value of these methods of assessment to food security.

It is clear that efficient food distribution is an important part of the food supply chain because it will provide safe food where it is required. The requirement of fit-for-purpose packaging and preservation of foods is necessary and a consideration for any product development. An understanding of how food businesses and transport business are clustered geographically is important when considering the structure of food supply. The resulting foodscapes can be used to plan sustainable distribution infrastructure that provides safe storage and transport of foods so that access to affordable foods is possible. In order

Table 4.4.   Analysis of current-state food and beverage distribution situation for SMEs and micro-companies with what could be achieved

What distribution situations currently exist	What could improve this situation, making food distribution more sustainable	Potential innovation intervention area
Own distribution resources used	Group distribution. Cooperate with trade members and use specialist haulage.	Route planning CRM with suppliers
Distribution cost is 10% of turnover	Implement new cost-saving technologies. Increase fuel and transport costs create need to implement cost saving technologies and networks	Biofuel utilisation Accounting for carbon dioxide emissions CRM with transport suppliers
Distribute nationally	Develop Internet and international retail. Impetus for internet marketing and international growth.	ICT applications and web solutions— selling and online book/reservation
Distribute less than 1 tonne of product daily	Cooperation between suppliers to rationalise high amounts of small load distribution.	Food groups and cooperative initiatives
Distribute ambient and chilled	Utilise freezing and other forms of preservation.	New materials and methods of preservation
Distribute using pallets	Utilise retail-ready, reusable, and recyclable packaging.	Designing out waste Life cycle analysis approaches
Own spare storage capacity	Cooperation between suppliers to optimise storage	Production and distribution planning/ scenario generation Design out waste

Source: The data were obtained from the research of W. Martindale and developed during the Flow Project (2010), which involved 60 food and beverage companies in the Yorkshire and Humber Region of the United Kingdom.

to demonstrate these principles, we will use the case study of regional food distribution that was developed in 2007 with the Regional Food Group in Yorkshire and Humber, UK. The project was called 'Flow', and it tackled the issues of food distribution in the regional food industry; the reasons were on the surface obvious: in 2007, the volatility of fuel price was all too apparent, and there was ominous policy pressure based on environmental (largely GHG emission focussed) credentials to increase road excise duty or introduce congestion charging in urban areas. These pressures have not disappeared and remain issues Flow has responded to, in the context of running a food business in Yorkshire and Humber, UK. It was clear that the transport of variable order volumes, energy, and fuel consumption pressures on the smaller food companies manufacturing perishable goods could be improved where there was duplication of delivery destinations, and this provided opportunities to strategically rationalise food logistics.

The Flow study of the food and beverage industry in Yorkshire and Humber collected data from 52 of the 220 Regional Food Group (RFG, a trade organisation) companies to provide an initial snapshot of how food and beverage products were transported, what volumes were transported, and what were the major issues facing transport of goods in RFG companies. The project findings are summarised in Table 4.4, and routes to improvement are highlighted. The food businesses of the region were also mapped using a Geographic Information System (GIS) to show where the location of different food sectors in the Yorkshire and Humber Region are. Generally, manufactured products will be associated with major transport routes and population (Figure 4.2). However, sectors such as dairy and meat that depend on rural production are more diffuse and distribute according to production centres (Figure 4.3).

The Flow study approach identified that the distribution of vegetable and fruit producers is associated with the

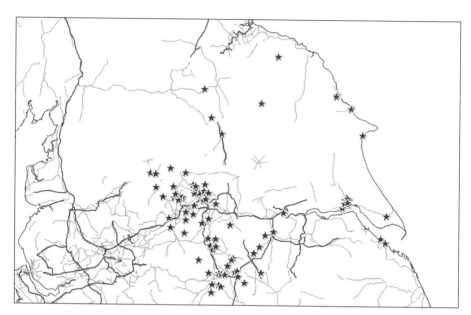

**Figure 4.2.   The Yorkshire and Humber (UK) bakery sector. The distri-
bution of bakeries is clustered in the Southwest of the region and is
associated with the areas of Leeds, Bradford, and Sheffield. The major
transport routes are of particular significance to this sector because
large amounts of transport occur in the mornings, and there is an
extremely quick transition from products leaving the manufacturing
lines to despatch into vehicles because of shorter product shelf life
and the shopper requirement for fresh products. The requirement for
storage space is thus lower, and there is likely to be spare storage
and vehicle capacity in the afternoons for bakery companies.**
Source: **Data from Ordnance Survey © Crown copyright (2008).**

production and growing regions for primary produce and
processing sites, such as trimming and washing plants.
This is associated with the best growing regions and more
productive soils suited for horticultural production. Manufacturers require good transport links for principally bulk
transport of fruit and vegetables that are perishable and
need to be delivered to customers or preservers in a timely
way. The distribution of confectionery manufacturers is
associated with the historical importance of confectionery
products. The confectionery sector will distribute at

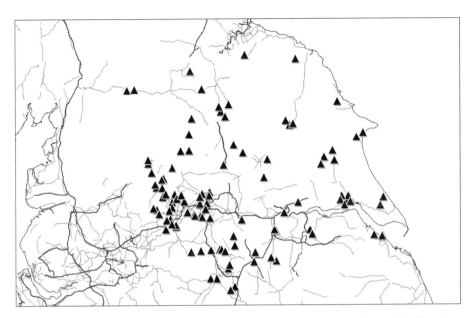

Figure 4.3.   The Yorkshire and Humber meat sector. The distribution of meat manufacturers is clustered towards the West of the region. This relates to the distribution of population in the Bradford, Leeds, Huddersfield, and Sheffield areas, and the centres of livestock production with supported services, including livestock markets and abattoirs. Most meat producers will have storage facilities for chilled and frozen products. Storage will often be at full capacity and despatch will nearly always be chilled or frozen.
Source: **Data from Ordnance Survey © Crown copyright (2008).**

ambient temperatures and not require chilled or frozen storage capacity. Beverage manufacturers are clustered towards the population centres. The distribution of dairies and ice cream manufacturers is diffuse, but there is a clear clustering towards areas of population because of the need for major transport routes for transporting fresh milk is of importance, and proximity to motorways is a trend in dairy distribution. Fat and oil manufacturers are associated with importing product through ports, as with the distribution of seafood manufacturers associated with landings of product.

The delivery destinations of the 52 Flow companies showed a clear clustering of delivery points within the Yorkshire and Humber region demonstrating that food distribution for small- and medium-sized companies remains relatively local, with most deliveries not being more than 50 km from point of dispatch and production. The study also enabled an assessment of the direct costs of operating trucks and vans to distribute products. A lorry with capacity for 26 pallets will result in a pallet space cost of £3 076 per year, increasing to £11 372 per year for a pallet space using a smaller 1 tonne payload van. A 7.5-tonne lorry will entail a £6 598 cost of a pallet space per year. The costs do not include fuel or tyres, and other variable fuel costs will clearly increase pallet space costs by between £1 400 (articulated lorry with 26 pallet spaces) and £3000 (van with one pallet space) a year. Developing studies that relate sustainability criteria to costs are essentially to delivering efficient supply chains because the sustainability criteria of a business and the business or financial systems in companies do not work together or integrate usually.

Data capture for food distribution systems is critical in enabling this type of approach, and it is supported by the emergence of open-source geo-information databases that can be utilised to obtain supply chain data. This is an important future development in the risk mapping arena for logistics. Table 4.5 uses this type of application, and the data sets used to convert the primary distribution data (volume of product transported and distance transported) are publically accessible. The data from the Flow study presented in Table 4.5 are an assessment of the destinations of food product deliveries within 70 km of Leeds City Centre, UK in a typical week for 10 meat product-manufacturing smaller companies in the Yorkshire and Humber region within 70 km of Leeds. The impact of distribution for these companies within the distance assessed as concentric circles from the centre is reported in the table.

Table 4.5. The delivery points for 10 meat product manufacturing companies located within 70 km of Leeds and their associated impacts within 70 km radius of Leeds in a typical week for 10 meat product-manufacturing companies in the Yorkshire and Humber Region

Radius (km)	Number of destinations	Kilometre travelled	Diesel consumed (L)	CO_2e emitted (kg)	CO_2e cost (£)	Total social cost (£)
10	5	374	196	526	1.66	102.95
20	2	90	147	393	0.40	24.82
30	2	157	47	125	0.70	43.23
40	6	270	86	231	1.19	74.17
50	3	129	40	107	0.57	35.55
60	4	265	94	251	1.17	72.88
70	4	114	101	271	0.50	31.22

Notes: Fuel consumed has been calculated using conversion constants for product freight described in research presented by Martindale and others (2008).[21] The fuel consumed for whole freighting operations (products and vehicles) have been obtained using the Department of Transport Freight Best Practice KPI publications. This was typically 3.6 km per litre of diesel for HGV and LGV vehicles. Conversion of fuel consumption to CO_2e amounts has been carried out using the method presented by Martindale and others (2008). The conversion factors for the economic cost of food transport including CO_2e (£0.0044/km), and total social cost, which includes accidents (£0.0312/km), congestion (£0.2226/km), transport infrastructure (£0.0008/km), noise (£0.0057/km), and air quality (£0.0101/km), have utilised conversions derived from the Defra 2005 report cited on the validity of utilising food miles and an indicator of sustainability. The conversion factors were obtained by utilising the costs of CO_2, accidents, congestion, transport infrastructure, noise and air quality reported by Defra in the United Kingdom and dividing them by the reported food miles by HGV (5.8 Bn km) and LGV (4.7 Bn km) vehicles to obtain the typical cost per km for a particular impact.[45] The sum of accidents, congestion, transport infrastructure, noise and air quality cost is presented as the sum of social cost.
Source: The data were obtained from the research of W. Martindale and developed during the Flow Project (2010), which involved 60 food and beverage companies in the Yorkshire and Humber region of the United Kingdom.

4.9 Developing Diets for Improved Sustainability and Health Criteria

We have now seen the need to obtain optimal micronutrient and protein nutrition so that good health is delivered from a balanced diet, and it is crucial to understand how these principles can help to deliver a sustainable diet. While the focus of food security has been largely on underconsumption of energy and protein, the converse food security issue is overconsumption of foods that has an increased impact on both food security and food supply. Increased overconsumption of foods will result in diseases associated with poor nutrition and increased food waste. In this context, it is crucial to understand why people overconsume food and how this interacts with the use of food in the home that results in increased waste. In order to demonstrate how analysis of food supply chains can provide insights into how consumers utilise food products, we use a case study here that defines the problems of overconsumption of food in the United Kingdom with respect to the retail landscape and food-scape of an area.

The United Kingdom has one of the highest levels of obesity in Europe, and research into the causes of obesity has suggested the development of obesogenic environments is responsible. Research carried out by the author of this book and Peter Kucher in Sheffield, UK has explored whether physical aspects of the environment and the retail landscape associated with food accessibility are responsible or not. The aim of our study has been to examine and compare areas of low and high children's obesity and the elements of physical environment in the county of South Yorkshire in the United Kingdom that might result in an increased prevalence of obesity. The results of the analysis suggest that there is no variation between areas of low and high percentage of obese children's body mass index (BMI) when we consider the number of food outlets and other geographic elements of the built and physical environment. Proximity to food retail outlets, restaurants, and, green and open spaces do not show any significant variation with respect to areas of high and low obesity. This contradicts the research into obesogenic environment, which suggests that built and physical environment have an effect on obesity.[46] The data for the study in South Yorkshire were obtained from the National Obesity Observatory (NOO) and the UK Office of National Statistics, where the proportion of obese children for specific areas can be identified.[47] The data sets provided by the National Obesity Observatory and Public Health England are very useful in giving the opportunity to develop powerful demonstrations of geographic analysis. Our initial studies have used geographical interpolation methods that combine the physical and built elements to solve specific policy challenges, such as improving public health, and by understanding these attributes associated with the physical environment, we can identify how they have an impact on specific health problems, such as obesity. Similar approaches can be used for other public health issues associated with food consumption, and these geospatial approaches may help us to define what is a sustainable diet by interpolating several health and sustainability criteria spatially.

We know that understanding nutritional requirements for health is essential if we are to communicate what sustainable diets are, and scientific studies have established what nutritional requirements are and what healthy meals should be. There are standard texts that provide data on the composition of foods and meals; the challenge facing us currently is to extend these data to dietary advice and recipes that the food industry can use for health and sustainability criteria.[48] Protein dominates sustainable consumption issues for diets, and the intake of essential amino acids is critical to obtaining sufficient protein in human nutrition, with World Health Organisation (WHO) guidelines stating that at least 0.66 g of protein per kilogram of body weight is required per day for maintenance of metabolism. Most importantly, lysine and the sulfur-containing amino acids (methionine and cysteine) will determine the nutritional quality of the protein source, and these attributes of protein nutrition defined by WHO can be placed into the context of recipes and diet. In terms of meals and diet, different food categories have variable protein value with respect to protein and ability to be used in recipes that deliver sufficient protein nutrition.

Fruits generally fall below WHO protein requirements but provide much of the micronutirients previously discussed, and root vegetables are mixed in meeting WHO protein requirements. Leafy/green vegetables (asparagus, broccoli, kale, spinach etc.) can provide important protein sources, but grains are a crucial source of dietary protein outside of livestock sources. Grains are poorer lysine sources, with the exception of oats. Nuts and seeds vary in protein content, with species like pumpkin providing protein-rich seeds. Legumes are a very good source of protein and dominated by the consumption of soy products, and as a protein food group, they are comparable to meat, eggs, and milk.

Table 4.6 shows the protein content and essential amino acid content for livestock products and define why they are good protein sources with respect to the already described food categories and vegetable products. These type of data are the kind that need to be formulated into planned diets according to the scientifically determined composition of diet that is communicated by WHO and other organisations. Protein is not the only point of challenge for the future, although the role of protein and micronutrients in regulating satiety holds much promise for managing overconsumption and obesity. A further challenge is to manage carbohydrate loading in diets because of diseases that result in poor regulation of carbohydrates, specifically the diabetes group of diseases, are at similar epidemic levels to obesity. Indeed, the relationship between

Table 4.6.   Protein content of livestock- and vegetable-derived foods: protein and amino acid contents of common foods (grams per 1000 cal)

	Grams per 1000 cal		
	Protein	Lysine	Methionine + cysteine
Animal products			
Beef	72.67	6.05	2.67
Cheddar cheese	61.79	5.14	1.93
Chicken	86.51	7.02	3.45
Egg	81.16	5.83	4.41
Lamb	63.22	5.58	2.38
Full fat milk	53.93	4.28	1.85
Vegetable products			
Broccoli	106.43	5.04	1.93
Kale	66.00	3.94	1.52
Spinach	129.13	7.91	3.91
Grain products			
Brown rice	23.24	0.89	0.80
Oats	43.42	1.80	1.85
Rye bread	32.82	0.90	1.20
Wholewheat bread	39.43	1.23	1.50
Wholewheat spaghetti	42.98	0.95	1.59
Legume products			
Lentils	77.76	5.43	1.68
Peas	66.91	3.91	1.41
Soya milk	83.33	5.42	2.64
Tofu	106.32	7.00	2.83

Sources: These data were adapted from the following sources: USDA National Nutrient Database for Standard Reference, available at http://ndb.nal.usda.gov/ (accessed 24 April 2014); the McCance and Widdowson's (eds) *The Composition of Foods*, 5th (1991) and 6th (2002) edn; the Diogenes GL Database MRC, available at http://www.mrc-hnr.cam.ac.uk/research/research-sections/nutrition-health-interventions/gi-database/ (accessed 1 May 2014); Foster-Powell, K., Holt, S. H. A., & Brand-Miller, J. C. (2002). International table of glycemic index and glycemic load values: 2002. *The American Journal of Clinical Nutrition*, 76(1), 5–56.

Table 4.7. Fibre, calories, and GI per 1000 cal for a range of foods

Food	Fibre (grams per 1000 cal)	Energy (calories per 100 g)	Glycaemic load per 1000 cal
Potatoes	20	85	276
Sugar, white	0	387	232
Rice, brown	10	110	210
Bread, wholemeal	28	250	180
Pasta, white	8	130	126

Source: These data were adapted from the following sources: USDA National Nutrient Database for Standard Reference, available at http://ndb.nal.usda.gov/ (accessed 24 April 2014); the McCance and Widdowson's (eds) *The Composition of Foods*, 5th (1991) and 6th (2002) edn; the Diogenes GL Database MRC, available at http://www.mrc-hnr.cam.ac.uk/research/research-sections/nutrition-health-interventions/gi-database/ (accessed 1 May 2014); Foster-Powell, K., Holt, S. H. A., & Brand-Miller, J. C. (2002). International table of glycemic index and glycemic load values: 2002. *The American Journal of Clinical Nutrition*, 76(1), 5–56.

them are the causal drivers of overconsumption, and improved dietary communication to consumers by the food industry is largely untested in the way that environmental impacts are with respect to the use of footprinting and LCA methodologies.

Table 4.7 shows the fibre, energy content, and glycaemic index for a number of carbohydrate rich foods. Understanding glycaemic index is crucial here because it enables the planning of diet with regard to slow and fast release of sugars, essentially how quickly complex sugars such as starch are digested, and the sugar derived is absorbed into the circulatory system. Thus, with protein, we have the metabolic requirement for growth and health as a determinant of sustainable meals, and for carbohydrates, we have the glyceamic index. If a diet is composed of sugars that are only absorbed quickly, that is, they have a high glycaemic index, then diabetes risk is increased. The use of fructose syrups in beverages is controversial in this context because overconsumption of beverages with high glycaemic indices has been associated with increased trends in obesity and diabetes diseases.[49]

Relating supply chain functions to diets and meals has proved to be a difficult process to achieve, and it has often been overlooked and left to the presentation of culinary methods by the public relations and media industry associated with chefs and restaurants. This is likely to change in future, as the need to conserve livestock products, reduce waste and prepare more healthy meals become important to sustainability and health policies. The CSIRO Total Wellbeing Diet (TWD), which has been referred to previously, is an example of this approach, and it has come from the Australian Government's science agency, CSIRO.[50] The diet has been developed at CSIRO's Clinical Research Unit in Adelaide, South Australia, and has led to the development of the higher protein, low-fat diet that is nutritious, facilitates sustainable weight loss, and is supported by scientific evidence. The TWD CSIRO book has extended scope to recipes and diet plans that aim to stop overconsumption of foods, which in turn has a very clear sustainability and food security impact in that diet planning will reduce overconsumption, reduce waste, and improve health. This type of approach may enable consumers to assess options in dietary change based on science in future.

References

1 Tilman, D., Balzer, C., Hill, J., & Befort, B. L. (2011). Global food demand and the sustainable intensification of agriculture. *Proceedings of the National Academy of Sciences, 108*(50), 20260–20264.

2 Gandhi, I. (1975). *A world without want.* Encyclopaedia *Britannica* book of the year 1975 (pp. 6–17). Chicago: Encyclopaedia Britannica.

3   Henderson, J. P. (2000). 'Political economy is a mere skeleton unless…': what can social economists learn from Charles Dickens? *Review of Social Economy, 58*(2), 141–151.

4   Rifkin, J., Howard, T., & Georgescu-Roegen, N. (1981). *Entropy: a new world view.* Toronto: Bantam Books.

5   Hawken, P., Lovins, A. B., & Lovins, L. H. (2010). *Natural capitalism: the next industrial revolution.* London: Earthscan.

6   Hertsgaard, M. (1999). *Earth odyssey: around the world in search of our environmental future.* New York: Random House Digital.

7   Ramessar, K., Peremarti, A., Gómez-Galera, S., Naqvi, S., Moralejo, M., Munoz, P., & Christou, P. (2007). Biosafety and risk assessment framework for selectable marker genes in transgenic crop plants: a case of the science not supporting the politics. *Transgenic Research, 16*(3), 261–280.

8   Schau, E. M., & Fet, A. M. (2008). LCA studies of food products as background for environmental product declarations. *The International Journal of Life Cycle Assessment, 13*(3), 255–264.

9   Andersson, K. (2000). LCA of food products and production systems. *The International Journal of Life Cycle Assessment, 5*(4), 239–248.

10 Andersson, K., Ohlsson, T., & Olsson, P. (1994). Life cycle assessment (LCA) of food products and production systems. *Trends in Food Science & Technology, 5*(5), 134–138.

11 Finnveden, G., Hauschild, M. Z., Ekvall, T., Guinee, J., Heijungs, R., Hellweg, S., & Suh, S. (2009). Recent developments in life cycle assessment. *Journal of Environmental Management, 91*(1), 1–21.

12 Roy, P., Nei, D., Orikasa, T., Xu, Q., Okadome, H., Nakamura, N., & Shiina, T. (2009). A review of life cycle assessment (LCA) on some food products. *Journal of Food Engineering, 90*(1), 1–10.

13 Martindale, W., Finnigan, T., & Needham, L. (2013). Current concepts and applied research in sustainable food processing. Chapter 2. In B. K. Tiwari, T. Norton, &

N. M. Holden (eds), *Sustainable food processing* (pp. 11–38). Chichester: John Wiley & Sons.

14 Jungbluth, N., Tietje, O., & Scholz, R. W. (2000). Food purchases: impacts from the consumers' point of view investigated with a modular LCA. *The International Journal of Life Cycle Assessment, 5*(3), 134–142.

15 British Standards Institute (2008). Guide to PAS 2050; 'How to assess the carbon footprint of goods and services. Specification for the assessment of the life cycle greenhouse gas emissions of goods and services'. London. http://shop.bsigroup.com/en/forms/PASs/PAS-2050/ (accessed 2 September 2013).

16 Dairy Co (2010). Milk roadmap. http://www.dairyco.org.uk/resources-library/technical-information/business-management/milk-roadmap/ (accessed 1 May 2014).

17 EBLEX (2012). Down to earth, the beef and sheep roadmap phase 3. http://www.eblex.org.uk/publications/corporate-publications/ (accessed 16 August 2013).

18 Küsters, J. (1999). Life cycle approach to nutrient and energy effi ciency in European Agriculture. Proceedings of the International Fertiliser Society No. 438, Cambridge UK. http://www.fertiliser-society.org/society-proceedings/authors—k/kuesters-j/proceeding-438/c-23/c-947/p-637 (accessed September 2, 2013).

19 Harvey, C. W., & Dutton, J. V. (1993). Root quality and processing. In D.A. Cooke & R. K. Scott (eds), *The sugar beet crop* (pp. 571–617). Dordrecht, The Netherlands: Springer.

20 Martindale, W. (2013). The sustainability of the sugar beet crop—the potential to add value. *British Sugar Beet Review, 81*, 49–52.

21 Sellahewa, J. N., & Martindale, W. (2010). The impact of food processing on the sustainability of the food supply chain. In W. Martindale (ed.), *Delivering food security with supply chain led innovations: understanding supply chains, providing food security, delivering choice aspects of applied biology* (p. 102). Warwick: Association of Applied Biologist. http://www.shu.ac.uk/_assets/pdf/foodinnov

-wm-impact-processing-sustainability-food-supply
-chain.pdf (accessed 2 May 2014).

22 Hülsbergen, K.-J., Feil, B., Biermann, S., Rathke, G.-W.,
Kalk, W.-D., & Diepenbrock, W. (2001). A method of
energy balancing in crop production and its application
in a long-term fertilizer trial. *Agriculture, Ecosystems and
Environment, 86*, 303–321. http://www.kunstmest.com/
files/energy08.pdf (accessed 1 May 2014). This paper
was the basis for several EFMA reports including 'Harvesting solar Energy using Fertilisers'.

23 Martindale, W., & Trewavas, A. (2008). Fuelling the 9
billion. *Nature Biotechnology, 26*, 1068–1070. http://
www.nature.com/nbt/journal/v26/n10/pdf/nbt1008
-1068.pdf (accessed 1 May 2014).

24 Smil, V. (1997). Global population and the nitrogen cycle.
Scientific American, 277(1), 76–81.

25 Smil, V. (1999). Detonator of the population explosion.
Nature, 400(6743), 415–415.

26 Draycott, P., & Martindale, W. (2000). Effective use of
nitrogen fertiliser. *British Sugar Beet Review, 68*, 18–21.

27 Sylvester-Bradley, R. (1993). Scope for more efficient use
of fertilizer nitrogen. *Soil Use and Management, 9*(3),
112–117.

28 Growhow UK Ltd (2013). CountingCarbon. http://
growhow.co.uk/content.output/444/444/Company
Information/CompanyInformation/CountingCarbon
.mspx (accessed 1 May 2014).

29 Wallén, A., Brandt, N., & Wennersten, R. (2004). Does the
Swedish consumer's choice of food influence greenhouse
gas emissions? *Environmental Science and Policy, 7*(4),
525–535.

30 Walmart (2012). The sustainability index. http://
corporate.walmart.com/global-responsibility/environ
ment-sustainability/sustainability-index (accessed 16
August 2013).

31 Martindale, W. (2014). Using consumer surveys to determine food sustainability. *British Food Journal, 116*(7), in
press.

32 Boland, M. J., Rae, A. N., Vereijken, J. M., Meuwissen, M.
P. M., Fischer, A. R. H., van Boekele, M. A. J. S.,

Rutherfurd, S. M., Gruppen, H., Moughan, P. J., & Hendriks, W. H. (2012). The future supply of animal-derived protein for human consumption. *Trends in Food Science and Technology*, 29(1), 62–73.

33 Defra (2012). Food statistics pocketbook. p.42. https://www.gov.uk/government/organisations/department-for-environment-food-rural-affairs/series/food-statistics-pocketbook (accessed 16 August 2013).

34 Friel, S., Dangour, A. D., Garnett, T., Lock, K., Chalabi, Z., Roberts, I., Butler, A., Butler, C. D., Waage, J., McMichael, A. J., & Haines, A. (2009). Public health benefits of strategies to reduce greenhouse gas emissions: food and agriculture. *The Lancet*, 374(9706), 2016–2025.

35 Quorn foods Ltd (2012). The mycoprotein web-site. http://www.mycoprotein.org/ (accessed 16 August 2013).

36 Macdiarmid, J., Kyle, J., Horgan, G., Loe, J., Fyfe, C., Johnstone, A., & McNeill, G. (2011). Livewell: a balance of healthy and sustainable food choices, Commissioned by WWF-UK. http://www.wwf.org.uk/wwf_articles.cfm?unewsid=4574 (accessed 16 August 2013).

37 Nielsen, P. H., Nielsen, A. M., Weidema, B. P., Frederiksen, R. H., Dalgaard, R., & Halberg, N. (2009). The LCA food database. http://www.lcafood.dk/ (accessed 16 August 2013).

38 Ridoutt, B. G., & Pfister, S. (2010). A revised approach to water footprinting to make transparent the impacts of consumption and production on global freshwater scarcity. *Global Environmental Change*, 20(1), 113–120.

39 Ridoutt, B. G., Eady, S. J., Sellahewa, J., Simons, L., & Bektash, R. (2009). Water footprinting at the product brand level: case study and future challenges. *Journal of Cleaner Production*, 17(13), 1228–1235.

40 Nielsen, P. H., Nielsen, A. M., Weidema, B. P., Frederiksen, R. H., Dalgaard, R., & Halberg, N. (2009). The LCA food database. http://www.lcafood.dk/ (accessed 1 May 2014).

41 FAO (2010). Greenhouse gas emissions from the dairy sector: a life cycle assessment.

42   Wiedemann, S. G., & McGahan, E. J. (2010). Environmental assessment of an egg production supply chain using life cycle assessment, final project report. A report for the Australian Egg Corporation Limited.

43   Ercsey-Ravasz, M., Toroczkai, Z., Lakner, Z., & Baranyi, J. (2012). Complexity of the international agro-food trade network and its impact on food safety. *PLoS ONE, 7*(5), e37810;   http://www.plosone.org/article/info:doi/10 .1371/journal.pone.0037810 (accessed 1 May 2014).

44   McLeod, K. S. (2000). Our sense of Snow: the myth of John Snow in medical geography. *Social Science & Medicine, 50*(7), 923–935.

45   Smith, A., Watkiss, P., Tweddle, G., McKinnon, A., Browne, M., Hunt, A., & Cross, S. (2005). The validity of food miles as an indicator of sustainable development-final report. Report ED50254.

46   Lake, A., & Townshend, T. (2006). Obesogenic environments: exploring the built and food environments. *The Journal of the Royal Society for the Promotion of Health, 126*(6), 262–267.

47   The National Child Measurement Programme (2013). Meta-Data download. http://www.noo.org.uk/NCMP (accessed 6 January 2014).

48   McCance, R. A., & Widdowson, E. M. (1960). The composition of foods. Medical Research Council Special Report Series 297 (3rd rev. edn of Special Report No. 235).

49   Popkin, B. M. (2006). Global nutrition dynamics: the world is shifting rapidly toward a diet linked with noncommunicable diseases. *The American Journal of Clinical Nutrition, 84*(2), 289–298.

50   Noakes, M., & Clifton, P. M. (2006). *The CSIRO total wellbeing diet*. Camberwell: Penguin.

5 Challenges and Solutions

5.1 The Food System Challenge of This Century: Is a Sustainable Diet Now Defined?

We have now seen that the determination of food product carbon footprint can enable the development of dietary planning across populations that result in reduced GHG emission impacts. This has been reported by Wallén, Brandt, and Wennersten (2002),[1] where sustainable food production projections for the Swedish national diet included large increases in potatoes, root vegetables and pulses (greater than 50%), and a reduction in sweets and soft drinks (greater than 50%) would result in sustainable health and environmental outcomes. Notable increases in fish and other cereals approaching 50% were also suggested by this research as a possible route to obtaining sustainable food consumption nationally.[1] This study, together with established data sets of life cycle assessment (LCA) information, provide important sources of information for the food industry that increasingly has the requirement to identify the sustainability criteria of the products supplied to customers. The Wallén, Brandt, and Wennersten[1] study also considers the proportional footprint of fossil energy used to produce a food product. Thus, there is a consideration of products that are not a large proportion of a diet but offer significant health benefits, and those which may use a lot of energy but much of this is not

Global Food Security and Supply, First Edition. Wayne Martindale.
© 2015 John Wiley & Sons, Ltd. Published 2015 by John Wiley & Sons, Ltd.

derived from fossil fuel, for example, the use of biofuels (recyclable energy) or nitrogenous fertilisers (non-recyclable energy) for forages and grazing.

Many commentators currently state we are some way from defining what a sustainable diet is and we are far from implementing sustainability policy that is relevant across an organisations supply chain activity of sourcing ingredients and retailing foods. However, there is a growing body of evidence that is providing, LCA data, GWPs, accredited footprints, and conversion factors for water use, energy use, GHG emissions, and other environmental attributes. With these conversion factors and LCA data, directors of supply chain activities can measure the GHG, water, and other attribute impacts of their supply functions. Communicating this in terms of a typical meal and diet has become possible, but there is still a requirement to communicate the complex science that has developed these methods to consumers so that they are able to assess consumption in terms of sustainability and security in a practical way. Of course, current LCAs will typically determine the specific sustainability criteria of products under investigation as a kilogram of product such as a 'kilo of beef' or a 'kilo of wheat grain'. They are not concerned with how these relate to food brand or retailer brand values. As the data concerning sustainability criteria become more open source and available to organisations who wish to define their supply chain sustainably, they will be able to embed sustainability criteria into business values. Clearly, the units of measurement that are typically in the form of mass of agricultural production are not particularly useful to retailers or manufacturers dealing with retailers because the ultimate functional unit a consumer deals with in a food supply chain is a recipe or a plate of food.

The interface between the manufacturer and retailer is critical here because as we have previously stated, food manufacturers are in a position to collate sustainability evidence relating to ingredients that are formulated into food products. This will ultimately add value to products for retailers who increasingly want to report sustainability criteria associated with branded products. Retailers will need to communicate sustainability criteria in terms of meals and diet to consumers because this is how we use food. We have seen this occur for health and well-being attributes of products in Europe for nearly 20 years now. It is time for sustainability to make a similar impact on the retail functions of the supply chain using scientific evidence to support it. The issue of traceability of products still presents major problems because we have uncertainty in the determination of where ingredients come from. Ultimately, regulation will strengthen the application of traceability procedures beyond accreditation. This will include the use the geographic information system (GIS)-based programmes that have been demonstrated here. These approaches can determine where products come from and associate the location of production with specific attributes, such as soil conservation, biodiversity, and GHG emission, wherever the sustainability analysis is made in the food supply chain.

It is important to understand why this situation has presented itself because it is almost blindingly obvious that the food industry is ultimately in the business of supplying meals to consumers, and those meals should be sustainable, safe, and nutritious. Therefore, why has the LCA arena been so dramatically diverted with food such that it reports products in terms of kilogram amounts or individual portion sizes at best instead of diets or meals? An important part of determining what the future sustainable plate of food is will be the impact of how retailers respond to sustainable drivers in the food industry. Current indicators suggest that retailers are responding to a number of issues that promote sustainability and corporate responsibility. This is certainly not unexpected because food retailers have been reporting environmental indicators for many years, and some have strived to obtain an overall sustainability index for all of their organisational efforts to develop sustainable food products. Notable systems have historically been based on providing assurance and traceability in supply chains. These are important attributes of any supply chain, but they are specifically important for food products because they are a prerequisite for providing robust food safety measures.

Assurance and traceability schemes developed by or with retailers have provided environmental reporting, established sustainability criteria for food supply chains, and changed supply chain practices. These include the following:

1. **LEAF Marque**: The Linking Environment and Farming systemic approach started in the 1990s and has remained under the leadership of Caroline Drummond as Chief Executive since it was established. The first LEAF audits applied an environmental management system (EMS) approach to assessing farm sustainability. The approach has been a stunning success, and it has extended to a retailer assurance scheme approach where the LEAF Marque label on food products is associated with a farm audit and sustainability criteria.[2]
2. **Marks and Spencer, Plan A** : There are several food retailer assurance schemes that use sustainability criteria. A notable scheme has been developed by the Marks and Spencer company, who have clearly linked assurance to sustainability criteria.
3. **Walmart, Sustainability Index**: This assurance scheme has linked sustainability with traceability across supply chains. This type of approach has exposed complexities, and the goal of obtaining an index that describes all sustainability actions across an organisation remains to be met.
4. **Roundtable on Sustainable Palm Oil (RSPO)**: This scheme is focused on a specific ingredient that is used

intensively by food and FMCG supply chains, palm oil. RSPO is a global, multi-stakeholder initiative on sustainable palm oil. The principal objective of the RSPO is 'to promote the growth and use of sustainable palm oil through co-operation within the supply chain and open dialogue between its stakeholders.' RSPO work to 'Principles and Criteria for Sustainable Palm Oil Production' that suppliers must follow. The certified sustainable palm oil (RSPO oil) is traceable through the supply chain by certification of each processing facility along the supply chain that processes or uses the certified oil. This is important because palm oil is produced in areas of the world where changes in land use associated with palm oil production can be related to the clearing of rainforests if RSPO standards are not met, and these changes would not be visible to consumers if RSPO standards did not exist (Figure 5.1). Palm oil represents another case of an ingredient that is utilised in many food products and FMCGs, assuring the supply chain for it is complex and requires a multi-stakeholder approach.

5. **Global Good Agricultural Practice (Global GAP)**: The value of assurance schemes in the food supply chain needs to be global, and the Global GAPs for agricultural products support this goal. The Global GAPs began in 1997 as EUREPGAP, an initiative by retailers belonging to the Euro-Retailer Produce Working Group. British retailers working together with supermarkets in continental Europe become aware of consumers' growing concerns regarding product safety, environmental impact, and the health, safety, and welfare of workers and animals. The Global GAPs provide a 'Chain of Custody' (CoC) Standard across the crop, livestock, and aquaculture product supply chain so that assurance and traceability are developed to enhance trust between functions of the supply chain

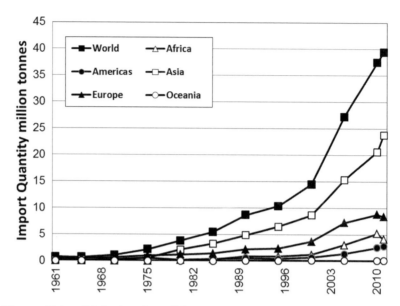

Figure 5.1. Global palm oil imports 1961–2012. The increasing palm oil imports are not only associated with biodiesel production, as significant increases have occurred before biofuels were promoted as liquid fuels that provided carbon neutral options. Palm oil is used to manufacture several household products, such as detergents and cosmetics, as well as food ingredients, and to provide a feedstock for biodiesel production.
Source: These data were adapted from FAO. FAOSTAT (2009). Trade, crops, and livestock products data set. http://faostat. fao.org/ (accessed 24 April 2014).

and due transparency is observed by consumers. The Global GAP CoC Standard identifies the status of products from farm to retailer.

6. **The Marine Stewardship Council (MSC)**: One of the first food assurance schemes developed, it is an independent non-profit organisation that provides sustainable fishing standards for fisheries. The MSC was founded in 1997 by the World Wide Fund for Nature (WWF) and the food manufacturer Unilever; it became fully independent in 1999. The MSC notably relate the

sustainable use of fishery resources to meal choices and provide a significant body of literature concerned with recipes and how to prepare fish and sea food produce. The MSC standard works to the principles of maintaining sustainable fish stocks in certified fisheries, minimising environmental impact of fishing operations, and using effective management by meeting local, national and international laws.

A significant theme for many schemes that result in labelling standards are the links of technical and sustainability information to meals and servings of food the consumer experiences. The Livewell diet developed with WWF has been developed to promote the consumption choices associated with a healthy and a sustainable diet. The Livewell example demonstrates this by promoting a diet that generally has 10–15% fewer livestock products and more vegetable protein associated with the meals it is made up of. Reducing livestock products in meals is likely to be a critical change in the look of future plates of food because the consumption of livestock products is extremely variable globally. Protein content of meals may remain at high levels in nations where livestock products are reduced, but the protein sources in a meal may change to include more vegetable proteins or industrially fermented proteins.

5.2 Supply Chain Challenges: Integrating the LCA Approaches in Agriculture, Manufacturing, and Retail

It is clear that a sustainable diet is part of the food security challenge if humankind is to deliver a food system sufficient for a projected 9 billion people in 2050 that uses lower energy inputs and produces lower waste within current land use limits.[3] Too often policy analysis of the food system recommends changes in agricultural production system and decreased consumption as the only means to meet this 2050 challenge.[4] This results in an oversight of supply chain functions in policy, where there is a requirement to assess impact from production through to consumption. This view is supported by research demonstrating consumption cannot be managed by consumer communication alone and supply chain solutions are urgently required.[5] Life cycle approaches to identify agricultural practices as a control point in supply must be integrated with food consumption. Established LCA methods provide assessments of the food system that have been overlooked in tackling the global challenge of doubling calorific and protein outputs. The LCA allows the definition of greenhouse gas (GHG) emission impacts as the global warming potential (GWP) and land use impacts, and it is time to use this thinking across supply chains.

Agricultural production successes will not continue to alleviate food supply chain stresses even though major food crops have achieved typical yield per unit area yield increases of 2–5% per year.[6] These trends are not a source of confidence even though increasing crop yields remain a valid target for policy development, because of the challenge of assessing ecosystem and food supply chain criteria with LCA methods.[7] The yield and ecosystem service components of agriculture are often analysed independently of the impact of genetic improvement and agronomic management, which has been shown to be an oversight in policy development.[8] Thus, an ecosystem service assessment approach to reporting yield is required, and LCA techniques provide this with regard to food supply chain functions. Ecosystem services have been considered part of the agricultural system since 1926, when Transeau first produced his treatise on embodied energy of maize crops when it was considered that the energy available to the agricultural system was boundless.[9] This systemic approach has been tested and refined, leading to accurate determination of farming energy balances and carbon footprint of European agriculture.[10] LCA methodologies are providing an option to policymakers who require a system-wide analysis of efficiency and ecosystem services associated with food supply chains and product life cycles.[11]

The life cycle approach not only identifies hot-spots of impact in supply chains, it can show where conservation of resources in supply chains is possible.[12] These develop trade-off relationships that would not be found using the yield assessment alone. These trade-offs are also important in the production of wetland rice, for example, where emissions of methane from production are traded with the maintenance of rice wetlands, which are an important global GHG sink.[13] Livestock systems provide a specific challenge to LCA methods because of the extreme diversity in production type, which includes grazing, housed, and integrated systems for animal husbandry. Despite variable production and methane emissions, the livestock production system can be defined by LCA, and the GWP or carbon footprint data for livestock products can be used to develop sustainable dietary scenarios. Animal feed conversion efficiencies to edible protein conversion efficiencies range from 5% for beef to 40% for milk in US production systems.[14] Thus, although livestock products make up 39.2% of global protein consumption, feed conversion efficiencies mean that LCA data are critical to develop policy that accounts for their ecosystem impact. The LCA approach shows beef production to be the more intensive user of land resources, with pork, chicken, eggs and milk having LCA-derived land use of less than 1 t/ ha. Grazing systems are critical to ecosystem services and global permanent pasture accounts for 3356×10^6 ha which is far greater than any other crop (FAOSTAT 2009 data).

The LCA approach identifies beef production as an important system for future improvement of both GWP and land use impact. Grazing systems are critical to the future development of the world food system, and LCA is beginning to account for livestock production impacts (see EBLEX[15] and Dairy Co[16] roadmaps). Ecosystem services associated with biodiversity in grazing systems are well tested, and there is further scope to assess those associated with soil and water management.[17] Thus, the efficient use of feeds and forages in grazing systems is critical to developing sustainable livestock production where it is well documented that the demand for livestock protein will increase.[18] This is why we must assess impact across food supply chains using LCA and FAO statistics, that show that the amount of livestock protein consumed as beef, pork, poultry, and fish globally is 43.80 million t/year, and average consumption trends for livestock protein have increased to 31g of livestock protein per day for each person globally (FAOSTAT data). Accounting for protein balance in this way hides huge variation and millions of individuals experience protein deficiency owing to a failure to optimise production, energy balance and waste across supply chains. LCA approaches will continue to provide insight into livestock supply chain functions, and it is the beginning of greater accountability that will be applied to other attributes, such as social responsibility, where LCA can be used to present changes in dietary consumption in supply chains. This requires linking LCA with census data so that impacts can be projected to population scale. The result must not be austere dietary policy because the culture of food and tastes are changing dramatically with the increased global production of garlic (37.3%), chillies (24.0%), and spinach (47.8%) in 2000–2010, being an indicator of this taste transition (FAOSTAT 2010 data). Taste and sensory factors are as important as production and will be a component of much future planning for the global food system.

In future, LCA application in policy development will map the spatial impact of consumption in urban environments. Martindale[19] has reported the mapping of food consumption in urban areas using GIS, LCA outputs, and census data. Examples of the scenarios developed include the consumption of a more sustainable diet containing 5% less meat and 10% more fresh fruit and vegetables as prescribed by World Wide Fund for Nature and the United Kingdom's Rowett Institute research (see the Livewell Diet)[20] is mapped across a population of a million urban residents of the South Yorkshire Region in the United Kingdom (Office of National Statistics, 2011).[21] The GWP data used for foods is now well tested and established, and the GIS-LCA hybrid approach we have developed allows the mapping of established LCA data with demographic data trends.[1] Thus, LCA can be used to measure the efficiency of production and consumption for individual products, supply chains, and urban food systems.

The scenario presented for South Yorkshire in the United Kingdom showed that around 0.9 million tonnes of GHG produced each year are associated with the dietary consumption of 1.14 million people.[22] This total figure relates to the UK national GHG inventory for the whole food supply chain of 61 million food consumers in the United Kingdom, which is reported to produce 178 million tonnes of GHG, where 50–60 million tonnes of the inventory is associated with consumption and waste (an additional 20 million t/year) activity.[23] The national GHG inventory approach provides a robust test for our GIS-LCA scenario, building a model using the UK national GHG inventory of 549 million tonnes of GHG.[24] The GIS-LCA approach is important because the sustainable criteria for supply chains vary spatially and the impact of consumption of products in a city will be dependant on demographic data, behavioural change, and population size.

5.3 Visualising the Data from the Food System Using GIS-LCA

Food supply systems are complex, with many attributes of food production integrating with waste, cooking, and safety behaviours that all require measurement if sensible policy is to guide sustainable development of food systems. While a difficult data acquisition task, GIS-LCA can map the attributes robustly according to the intensity of consumption space. The use of LCA is crucial to account for food supply chain impacts and to develop trade-offs between production, consumption, and impact so that sustainable options can be derived. There is a requirement for policy to interface consumption trends with those of taste so that future impacts such as non-meat to meat and other transitions in populations can be identified earlier and GIS-LCA-based methods can make projections of environmental impact. There is also a requirement to communicate trade-offs and mapping consumption data; providing mapped LCA scenarios represents an opportunity to plan policy for urban populations and their relationships with food supply chains. Mapping methods will also support the international accreditation systems for carbon labelling of individual food products.[25] While these accreditations are important sources of standardisation, mapping the impact of food consumption by populations provides an important future communication of carbon footprinting and LCA methodology in sustainability policy.

The food system includes producers, manufacturers, distributors, retailers, and consumers. While this segmentation model of the food system is straightforward, it is composed of several supply chains, and the number of these makes projecting the functioning of the food system incredibly complex. As an example of these complexities of scale, the European food system serves some 480 million people each day with safe and nutritious food and drink.[26] The food system is not just defined by volumes of transactions, because the current food needs of customers are becoming more complex, with environmental impact, social responsibility, functional foods, nutraceuticals, obesity, and food miles, among many other issues, driving new products and improved business development.[27] These factors are continually stimulating the demand for innovation in the food system.[28] Consumer purchase choices are increasingly associated with health, ethical, labelling, and environmental concerns; they are balanced with more traditional choices concerned with product variety, convenience, and out-of-home consumption. A representation of the food system can be provided by surveys that provide a complex continuum of consumer intentions and purchase reality.[29] For example, consumer intentions to buy ethical products are often in conflict with requirements of economical price and wide product choice. Developing communications for improved shopper understanding of the resource and social responsibility history (and potential future) of products will be of distinct competitive advantage in the food and beverage retail environment. Purchase decisions currently sum up these choices and can often show distinct gaps in consumer knowledge of food supply chains.[30] Identifying knowledge gaps in supply information will provide opportunities to implement more efficient consumer communications that are able to influence decisions for purchase based on evidence. This is critical because understanding supply chains can clarify environmental impact and 'food miles' issues associated with food and drink products.[31] The impacts of these issues in the food system have been defined by a number of models,[32] and the successful implementation of food supply chain sustainability and innovation will follow three goals; these are highlighted as the following:

1. The ability to utilise new and established technologies in the agri-food arena.

2. Sensing the regulatory and business environment to deliver products that optimise the use of energy in the supply chain.
3. Innovative multidisciplinary communications that provide smarter communications for consumers with health and ethical information about the food and drink they consume.

Health- and well-being-focused innovations have transformed how consumers purchase foods, and global production statistics are beginning to show emergent trends that focus on consumer demands for quality, nutrition, taste, and pleasure.[33] A broad example of this is evident in global production increases of agricultural products that have already been highlighted for taste transitions with spinach (337%), garlic (230%), and chillies (223%), but noted here for between 1990 and 2005. They suggest a dramatic change in the requirement of not only calories and protein, but also flavour, nutrition, and taste. Food staples, such as wheat (2%) and sugar crops (21%), have more conserved production increases over the same time period.

Furthermore, staple ingredients, such as sugar, provide a case example of how information regarding health, ethics, and environment can change consumer perception of products. The European sugar manufacturing and retailing sectors have experienced significant change because of ethical purchasing responding to differences in the European sugar production framework and the world production of cane sugar. The issues associated with sugar trade have been largely unknown to the consumer until issues such as 'Everything but Arms'[34] and the emergence of organic sugar (from cane and beet) became of increased importance in purchase choice. Furthermore, consumer awareness that overconsumption of sugar is often linked to health and obesity issues and it has produced considerable activities resulting in the increased use of slower release sugars (lower glycaemic indices), lower calorie sweeteners, and 'balanced' consumer nutrition policies.[35] It is clear health will continue to drive much innovation because it is clearly associated with enhanced pleasure and quality of life. We are also currently seeing the palm oil sector reviewing corporate social responsibility (CSR) systems in supply chains in response to consumer-led pressures. In a similar way, the coffee supply chain has recently experienced high-profile scrutiny of CSR issues.[36]

5.4 Technology Enablers and Opportunities

Developing supply chain approaches to stimulate food innovation requires the analysis of product life cycle from the farm to the fork (or more accurately, the farm to the home). Ultimately, supply is determined by activities that are pre-farm gate in supply chains. The food system is dynamic, and consumer concerns change according to the availability of food. For example, rapid changes in the distribution of food and beverage processing infrastructure have accompanied improvements in primary production on farms.[37] The limits in regional agricultural product supply have been ameliorated by efficient logistical infrastructure, consumer demand, preservation, and packaging of food.[38] These developments have hidden the full cost of not producing food regionally for consumers. We are now beginning to account for these limits in agronomic capacity with the emergence of assurance and environmental labelling schemes for food.[39] Current food security concerns suggest increased synergy between agriculture, food manufacture, and novel technologies are required in the future food system. This is becoming evident by the emergence of technologies for recipe-ready (eliminating processing and preservation),[40,41] lower allergenic,[42] and more efficient food materials[43] for production. Such technologies have a central role in closing knowledge gaps and increasing nutritive quality and subsequently the health of billions of people. Retailer, consumer, and regulator acceptance of novel technologies will be critical to the market entry of innovative technologies. Indeed, sensing the regulatory environment in the first instance will be essential to the successful market entry of innovative and novel technologies. Further understanding of how shoppers perceive food availability will be an important driver to responding to consumer concerns that surround novel technologies.

Communication portals and databases that provide research and evidence resources for food innovation, nutrition, environment, and health issues will be critical to the development of consumer communications. There is a requirement for communications to be framed within a theme, such as ethical purchase and need for the means of framing to be based on evidence. Ethical values associated with food products are closely related to issues of health for the consumer and CSR for food companies. For example, there is no doubt poor nutrition is a major cause of ill health and premature death in many developing and developed countries. The 'obesity epidemic' seems in conflict with the increased use of functional foods in diets and the emergence of health- and well-being-led innovations in the food and beverage industry. Such situations have tiers of the regulatory, policy, and research expertise in the public health arena thoroughly confused about how balanced diet and nutrition should or could be communicated to consumers. Clearly, the interactions between lifestyle, diet, and public health issues are not as simple as many commentators have thought, and understanding the obesity epidemic in emerging economies and developed nations is not straightforward. Communicating the importance of balanced diets is clearly a cornerstone of robust public health communication.

New recipe development and new product development (NPD) is a significant area for much innovation. Innovations need not be new and may solve long-term issues that relate to product quality in manufacture, cooking, storage, and distribution, and subsequently result in customer experience. As an example, regulator and allergenicity issues are currently driving innovative implementation of modified atmosphere packaging solutions. For example, the preservation of fresh produce without using sulfur dioxide and sulfites as preservatives has provided significant impact in the fresh produce sectors.[44] Sulfites are used as preservatives and have been associated with allergenicity; regulatory changes have moved to remove sulfites from foods. This has led to manufacturers considering the use of packaging solutions and modified atmosphere packaging that excludes oxygen. The modified atmosphere packing inhibits the action of phenyalanine lyase enzymes that cause browning of fresh foods in oxygen-containing atmospheres.

Thus, companies must integrate current food security policy initiatives into process and manufacturing design and guide many future activities that deliver sustainability and drive future policy. Companies can develop CSR systems that provide fairness by aligning their business operations with policy guidance and use traceability data to define the social responsibility associated with products. Responsibility will not only include lowering environmental impacts, but also consider human rights, welfare, and nutritional outcomes. Responsibility associated with nutrition is a current policy focus because of the rising levels of obesity in developed and developing nations.

The World Wild Fund for Nature has developed the Livewell Diet with the Rowett Institute, which has been highlighted previously, and the approach taken will become an important source of change in the food system. The diet is based on sustainable criteria and independent research carried out by the author shows that the 7-day diet has a reduced carbon footprint based on data derived from Wallén, Brandt, and Wennersten.[1] The reduction in carbon footprint as compared with the current diet in the United Kingdom reported by the National Diet and Nutrition Survey is between 5% and 10% lower per person per day. These represent small lifestyle changes that processors and manufacturers can stimulate using portion and product design options. However, initial research shows the waste food generated from more sustainable diets is greater than current food waste. These scenarios place important considerations on planning for sustainability. It is critical that sustainability planning by the processing and manufacturing sectors is integrated with changes in the way in which consumers utilise and dispose of food. This again represents opportunities for processors to consider the value of preservation and freezing options to enhance the sustainability criteria of food products.

The process of benchmarking sustainability criteria across food supply chains and populations will provide applications for food processors and manufacturers because it is clear that the sustainability of a supply chain cannot be assessed by carbon equivalent emissions, water use, or waste production alone. A far broader assessment of sustainability is required that relates social responsibility, carbon, water, and waste to the consumption behaviour of individuals and populations. Many farm and food assurance schemes attempt to consolidate sustainability criteria for products, but there are still significant opportunities to assess and communicate sustainability criteria for individuals and benchmark these to typical population trends.

The development of new products that target the healthiness and well-being markets has increased significantly since the 1990s when the UK government's Foresight group for food and non-food crops supply chain management identified health as being a major factor in the future food industry. However, the link to whole meals was not robustly made and left to recipe publications and popularisation of culinary skills and food preparation. This scant attention to whole meals and diets by the processing and manufacturing industry will not provide sustainable outcomes for food supply and needs to change. There have been significant areas of recipe development that have impacted on the food manufacturing and processing sectors. For example, traditional recipe planning used by manufacturers in NPD has made use of cheaper ingredients, such as fat and salt, to reduce the unit cost of product. However, health policy developments, such as the UK Food Standards Agency 'Five-a-Day' programme, have changed the approach of many manufacturers to NPD. For example, in the United Kingdom, commercial and government-led policy pressures have resulted to an increase in the utilisation of vegetables for 'bulking-out' (and cost reduction), the attainment of product marketing claims based upon the 'Five-a-Day' initiative, and a decrease in the salt content of recipes.

The reduction of salt has been area of much process innovation even though the impetus for salt reduction has been guided by regulators raising the issue of who drives the requirement for changes in recipes so that they align with healthier outcomes. The UK Food Standards Agency emphasised the requirement to reduce salt consumption to 6 g of salt or 2.40 g of sodium per adult per day. The UK government has reported that in 2008, sodium intake, excluding table salt and allowing 10% for wastage, was estimated to be an average of 2.78 g/person/day from household purchases plus food eaten outside the home. This is a reduction of 2.2% on 2007 and a 14% decrease since 2001–2002. Thus, the food manufacturing and processing sector has responded to a national health indicator. This is emphasised by the use of umami flavours that have been derived from seaweed or yeast products as alternative flavour enhancers and have provided innovative product formulation in the food processing sectors. These scenarios need to be integrated into whole meals for sustainable health outcomes.

Consumer data regarding the use of foods in households represents a distinct gap in our understanding of how consumers interact with foods and data on the experience of taste and satiety are an essential part of developing new products and changing how they are manufactured. Flavour and fragrance composition is recognised as a highly complex and technical aspect of formulating foods, but our understanding of how satiety is controlled will impact on how we manage the planning of resulting food product design for sustainable processing. This is an area where nanotechnology is providing new understanding of encapsulation and delivery of nutrients in the body. Understanding consumer experience of taste will provide insights into how nutrition and satiety interacts with consumption. The food industry can respond with novel approaches of formulating products; as an example, that use optimised micronutrient content in foods as a potential means of reducing consumption of energy-dense and nutritionally poor foods that enhance appetite and overeating. The opportunities for understanding taste, satiety, and consumer behaviour should be considered together with those of materials and the texture of foods by processors for innovative outcomes. These criteria are currently being considered with regard to products across the food industry to further develop CSR strategies.

References

1 Wallén, A., Brandt, N., & Wennersten, R. (2004). Does the Swedish consumer's choice of food influence greenhouse gas emissions? *Environmental Science and Policy,* 7(4), 525–535.
2 Lewis, K. A., Green, A., Tzilivakis, J., & Warner, D. J. (2010). The contribution of UK farm assurance schemes towards desirable environmental policy outcomes. *International Journal of Agricultural Sustainability,* 8(4), 237–249.
3 Godfray, H. C. J., Beddington, J. R., Crute, I. R., Haddad, L., Lawrence, D., Muir, J. F., Pretty, J., Robinson, S., Thomas, S. M., & Toulmin, C. (2010). Food security: the challenge of feeding 9 billion people. *Science,* 327(5967), 812–818.

4   de Boer, J., Hoogland, C. T., & Boersema, J. (2007). Towards more sustainable food choices: value priorities and motivational orientations. *Journal of Food Quality and Preference, 18,* 985–996.

5   Hoogland, C. T., de Boer, J., & Boersema, J. J. (2007). Food and sustainability: do consumers recognize, understand and value on-package information on production standards? *Appetite, 49,* 47–57.

6   Evans, L. T. (1998). *Feeding the ten billion.* Cambridge, UK: Cambridge University Press.

7   Lobell, D. B., Cassman, K. G., & Field, C. B. (2009). Crop yield gaps: their importance, magnitudes, and causes. *Annual Review of Environmental Resources, 34,* 179–192.

8   Tilman, D., Balzer, C., Hill, J., & Befort, B. L. (2011). Global food demand and the sustainable intensification of agriculture. *Proceedings of the National Academy of Sciences, USA, 108*(50), 20260–20264.

9   Transeau, E. N. (1926). The accumulation of energy by plants. *The Ohio Journal of Science, 26*(1), 1–10.

10 Brentrup, F., Küsters, J., Kuhlmann, H., & Lammel, J. (2004). Environmental impact assessment of agricultural production systems using the life cycle assessment methodology I. Theoretical concept of a LCA method tailored to crop production. *European Journal of Agronomy, 20,* 247–264.

11 Costanza, R., d'Arge, R., de Groot, R., Farber, S., Grasso, M., Hannon, B., Limburg, K., Naeem, S., O'Neil, R. V., Paruelo, J., Raskin, R. G., Sutton, P., & van den Belt, M. (1997). The value of the world's ecosystem services and natural capital. *Nature, 387,* 253–260.

12 Ridoutt, B. G., Eady, S. J., Sellahewa, J., Simons, L., & Bektash, R. (2009). Water footprinting at the product brand level: case study and future challenges. *Journal of Cleaner Production, 17,* 1228–1235.

13 Burney, J. A., Davis, S. J., & Lobell, D. B. (2010). Greenhouse gas mitigation by agricultural intensification. *Proceedings of the National Academy of Sciences, USA, 107*(26), 12052–12057.

14 Smil, V. (2002). Nitrogen and food production: proteins for human diets. *Ambio, 31,* 126–131. Royal Swedish Academy of Sciences 2002.

15   English Beef and Lamb Executive (2012). Down to earth, the beef and sheep roadmap phase 3. http://www.eblex.org.uk/publications/corporate-publications/ (accessed on 6 May 2014).

16   Dairy Co. (2010). Dairy roadmap. http://www.dairyco.org.uk/resources-library/technical-information/business-management/milk-roadmap/ (accessed on 23 April 2014).

17   Zaks, D. P. M., Barford, C. C., Ramankutty, N., & Foley, J. A. (2009). Producer and consumer responsibility for greenhouse gas emissions from agricultural production—a perspective from the Brazilian Amazon. *Environmental Research Letters, 4*, 1–12.

18   FAO (2006). *Livestock's long shadow: environmental issues and options.* Rome: Food and Agriculture Organisation.

19   Martindale, W. (2010). Food supply chain innovations. In W. Martindale (ed.), *Delivering food security with supply chain led innovations: understanding supply chains, providing food security, delivering choice aspects of applied biology* (pp. 1–6). Aspects of Applied Biology Vol. 102. Warwick: Association of Applied Biologists.

20   Macdiarmid, J., Kyle, J., Horgan, G., Loe, J., Fyfe, C., Johnstone, A., & McNeill, G. (2011). Livewell: a balance of healthy and sustainable food choices. Commissioned by WWF-UK.

21   ONS (2011). 2011 census, population and household estimates for England and Wales. http://www.ons.gov.uk/ons/rel/census/2011-census/population-and-household-estimates-for-england-and-wales/index.html (accessed on 23 April 2014).

22   Martindale, W. (2014). Mapping consumption, integrating LCA and GIS—it is possible and we have proven application. Blog including research reports available at http://waynemartindale.com/2014/03/27/mapping-consumption-integrating-lca-and-gis-it-is-possible-and-we-have-proven-application/ (accessed 5 June 2014).

23   Garnett, T. (2008). Cooking up a storm: food, greenhouse gas emissions and our changing climate. Food Climate Research Network. http://www.fcrn.org.uk/sites/

default/files/CuaS_Summary_web.pdf (accessed on 23 April 2014).

24 ONS (2012). Provisional 2011 results for UK greenhouse gas emissions and progress towards targets, last updated 29 March 2012. http://www.decc.gov.uk/assets/decc/ 11/stats/climate-change/4817-2011-uk-greenhouse-gas -emissions-provisional-figur.pdf (accessed on 23 April 2014).

25 British Standards Institute (2008). Guide to PAS 2050; how to assess the carbon footprint of goods and services. Specification for the assessment of the life cycle green- house gas emissions of goods and services. London. http://shop.bsigroup.com/en/forms/PASs/PAS-2050/ (accessed on 6 May 2014)

26 Raspor, P., McKenna, .B., Lelieveld, H., & de Vries, H. S. M. (2007). Food processing: food quality, food safety, technology in ESF-COST (2007) Forward Look: Euro- pean Food Systems in a Changing World.

27 UK Cabinet Office Strategy Unit (2008). Food: an analy- sis of the issues, discussion paper. http://webarchive .nationalarchives.gov.uk/+/http:/www.cabinetoffice .gov.uk/media/cabinetoffice/strategy/assets/food/ food_analysis.pdf (accessed on 6 May 2014).

28 Costa, A. I. A., & Jongen, W. M. F. (2006). New insights into consumer-led food product development. *Trends in Food Science & Technology, 17*, 457–465.

29 Corsten, D., & Gruen, T. (2003). Desperately seeking shelf availability: an examination of the extent, the causes, and the efforts to address retail out-of-stocks. *International Journal of Retail and Distribution Management, 31*, 605–617.

30 Schröder, M. J. A., & McEachern, M. G. (2005). Fast foods and ethical consumer value: a focus on McDonald's and KFC. *British Food Journal, 107*(4), 212–224.

31 Food Innovation, Sheffield Hallam University (2008). Platform presentation to Association of Applied Biolo- gist. Conference poster. February 2008. http://www .foodinnovation.org.uk/download/files/envhealth choice.pdf (accessed on 23 April 2014).

32 Linnemann, A. R., Benner, M., Verkerk, R., & van Boekel, A. J. S. (2006). Consumer-driven food product development. *Trends in Food Science and Technology, 17*, 184–190.
33 FAO STAT (2008). Crop production data sets. http://www.fao.org/corp/statistics/en/ (accessed on 23 April 2014).
34 European Commission (2014). EU and WTO trade policy updates. http://ec.europa.eu/trade/issues/global/gsp/eba/index_en.htm (accessed on 23 April 2014).
35 Foresight (2007). Tackling obesities: future choices report. http://www.foresight.gov.uk/Obesity/Obesity_final/Index.html (accessed on 23 April 2014).
36 Black Gold (2005). Directed by M. Francis & , N. Francis. http://www.blackgoldmovie.com/ (accessed on 23 April 2014).
37 Trewavas, A. (2002). Malthus foiled again and again. *Nature, 418*, 668–670.
38 Kumar, S. (2008). A study of the supermarket industry and its growing logistics capabilities. *International Journal of Retail & Distribution Management, 36*(3), 192–211.
39 Clements, M. D., Lazo, R. M., & Martin, S. K. (2008). Relationship connectors in NZ fresh produce supply chains. *British Food Journal, 110*(4/5), 346–360.
40 Morris, J., Hawthorne, K. M., Hotze, T., Abrams, S. A., & Hirsch, K. D. (2008). Nutritional impact of elevated calcium transport activity in carrots. *PNAS, 105*(5), 1431–1435.
41 White, P. J., & Broadley, M. R. (2005). Adding nutritional value to food ingredients by biofortication. *Trends in Plant Science, 10*(12), 586–593.
42 Singh, M. B., & Bhalla, P. L. (2008). Genetic engineering for removing food allergens from plants. *Trends in Plant Science, 13*(6), 257–260.
43 Graveland-Bikker, J. F., & de Kruifa, C. G. (2006). Unique milk protein based nanotubes: food and nanotechnology meet. *Trends in Food Science & Technology, 17*, 196–203.
44 The Food Manufacturer (2007). Northern Foods finds a way to stop spuds browning. http://www.foodmanufacture.co.uk/Business-News/Northern-Foods-finds-way-to-stop-spuds-browning (accessed on 6 May 2014).

6 The Future and Our Conclusion

6.1 The Future Food System

Visionaries in the corporate world who strive for their businesses to meet sustainability goals tend to associate them with risk reduction, while those who do not implement sustainable management tend to be deterred by risk. What should be always considered is that projections are nearly always wrong, in that they are projections, and this goes for population projections and business projections. The longer the time period the projections run for, the more likely they are to deviate; these basic rules are important to consider because projections may only be viewed as guides to future trends. Risks in business are all too often associated with failure, and herein are the problems with meeting sustainable goals. These are illustrated by Yvon Chouinard, CEO of the Patagonia Company, who determined human values of enjoyment and focused on components of a whole system to achieve sustainable outcomes.[1] These are achieved by balancing the production capacity to consumer requirements and the consumption impacts with the available options of recycling, reusing, and repairing products.[2] Measurements of success are classified in a dualistic way using quantitative and qualitative analysis that is now described, and the goal is to use both forms of analysis to reduce risks of implementing sustainability in businesses.

Global Food Security and Supply, First Edition. Wayne Martindale.
© 2015 John Wiley & Sons, Ltd. Published 2015 by John Wiley & Sons, Ltd.

Technical systems, such as LCA, that quantify sustainability attributes, such as embodied greenhouse gas emissions, enable labelling and evidence-based statements to be made regarding products. The technical measurement systems can also redirect strategy because LCA approaches can identify processes that result in lower embodied energy during a product lifecycle. In the case of Patagonia, LCA was utilised to show that polyester fibres were less impacting than cotton fibres with regard to embodied energy, greenhouse gas emissions, and water use. This measurement guided Patagonia in making design and resource decisions for the future. However, LCA and labelling activities of a business can only go so far in providing a measure of sustainability. They cannot form the only basis for an overall scorecard of sustainability.

Qualitative values within organisations are critical to attaining sustainable goals and measuring these presents numerous problems. The approach of visionary CEOs to date has been to develop integrated or step-by-step approaches to achieving an overall goal. The steps are based on the human values of enjoyment, fun, and respect. Results are impressive; such approaches work and trust is a core value across the supply chain and organisations involved with it. These caveats are explained in greater detail in Yvon Chouinard's book, *Let My People Go Surfing*, which notable ecological researchers have lauded as a handbook for responsible corporations. Approaches to sustainable business need to be bold, and they are exceptionally risky because they will move away from standard practices. What is critical is the perceived value of trust within an organisation. Valuing and quantifying trust may not be an impossible or complex task. Evolutionary biologists have quantified altruistic and selfish behaviours; indeed, these methodologies have transformed modern biology. Scientists can even quantify altruistic and selfish traits to specific genetic attributes. Behavioural models of altruistic nature have been applied in the financial sectors, but we have yet to test them within the sustainability arena. I believe that business leaders have developed methods of achieving sustainability through trust, and scientists now have an opportunity to formalise these trust values using evolutionary behavioural models.

An awareness of data capture and utilisation is critical; it will continue to change the way we work, and it will be the lifeblood of the new models of supply. The volumes of data required to predict and understand consumption are becoming attainable as a consequence of open sources, wireless communications, and online digital arenas. These represent an important development in humankind's ability to utilise data in the last decade. The data concerning consumption of products in populations provide important insights into how impacts are manifested in society. Corporates and consumers in the future will demand a requirement to measure the impact of consumption in populations with regard to their social, economic, and environmental values and attributes. It is impossible to appraise the sustainability attributes without making and assessment of supply chains for producing and consuming products.

Traditionally, supply chain data have been processed for labelling schemes that prescribe particular qualities on environment or fairness to a specific product. For producers, this has been delivered as the stewardship and assurance schemes; for manufacturers, it has been focused on health and safety, and retailers have provided a melting pot where each of these labels have been presented to the consumer to see what is perceived as most important by them. There have been notable successes, but it is a chaotic system of assessment that is not suitable to the future supply chain function where data are collected in real time all the time. The failures of the labelling system for sustainability are illustrated strongly by corporations led by visionary goals of creating a sustainability index for their organisation and products. When this has been attempted, it has failed but is it possible to do?

The increased ability to capture production and consumption data can be integrated with supply chain analysis methods to provide opportunity for the future corporate to exert decisions associated with measureable trust and fairness. The producers, manufacturers, retailers, and consumers segments of supply chains have traditionally had defined start and stop roles or points, but open-source information has made this type of segmented approach increasingly integrated with a more comprehensive data system. Models detailing the linear relationships of supply chains are not sufficient anymore, and there are opportunities to apply more dynamic theories of supply and consumption that transform our view of what is achievable and sustainable. Control theory offers one such methodology that is likely to be more applicable to our future world view because it accounts for supply chains working together and sums the impacts of components within a supply system. Control theory presents supply chains as branches in throughput pathways, and the flow of materials through them are proportional control coefficients. Control coefficients that are removed from equilibrium points are control points of the system. The sum of control of these pathways is always unity, and the control points are critical to the working of these pathways. Thus, control theory provides the means to analyse the operations effectively, such as supply and consumption, providing a means to measure them at specific points or definite parts of a system where they might vary in control; these are the critical control points. Measures of sustainability in systems and supply chains will focus on control points of supply and consumption; models based on control theory can be applied to determine where an intervention to a supply chain is likely to be most effective at improving the sustainability criteria.

A traditional view of where supply chains should be deployed is based on standard compound interest (return on investment) and maximum sustainable yield models. These approaches need to be made more robust and control theory provides a significant basis to determine how supply chains work; models of colonisation can also provide insights into where opportunities that are sustainable actually are. The models concerned with island biogeography can be applied and can integrate existing views of ecosystem service values. Current theories regarding connectivity can provide assessment of how robust supply and consumption dynamics are likely to be realised.

The threat of limited food security has been highlighted globally by the perfect storm scenario of recent years, where the attributes of environmental degradation, economic growth, population increase, and climate change have uniquely impacted on the world food system. This has focused intense policy activity on sustainable production, processing, and manufacturing of food products. The major issue we have not fully considered for food sustainability is managing responsible food consumption because it has a most important role to play in determining the environmental impacts of resource use as identified by key policy and research reports. Indeed, reporting of limitations within the food system has highlighted crisis situations for governments and the populations of nations. If we are to manage food consumption sustainably, it is necessary to investigate the resource flows across food supply chains where the processing and manufacturing functions have a critical role in delivering sustainable products. The role of processors and manufacturers will be critical because profitable business practices can be maintained by sustainably reducing resource use.

Food processors and manufacturers have a critical role in the supply chain with regard to designing products that are integrated into sustainable diets. Currently, measured sustainability criteria, such as the carbon footprint, are focused on the embodied energy and GHG emissions associated with individual products. However, populations eat whole meals, not individual products, and the sustainability criteria of diets are largely unknown and untested. I consider that the delivery of sustainable diets presents future opportunities to processors and manufacturers. This is because product development innovations can focus on the delivery of whole meals, and the associated GHG, water, and waste impacts are associated with diets where they have previously been reported for individual food products, not meals. Indeed, this presents a novel approach that many processors and manufacturers have not fully considered yet even though the food industry has developed meal guides, recipes, and recipe literature associated with specific products delivering nutritional goals for consumers. However, sustainable goals or impacts of food for consumers are not generally promoted in dietary guides and literature. This results in sustainability being perceived as an interesting but immeasurable consumption goal for many organisations and individuals. It does not have to be like this because the industry has the tools and skills that have been applied to improving nutritional communications that can provide a model for the promotion of sustainability.

Our analysis has provided the position that if consumers could measure and respond to the sustainability value of whole meals and their diet, then they would potentially change how they purchase and consume products for sustainable outcomes. Indeed, we suggest that the future consumer will hold both nutrition and sustainability criteria of foods with equal value because the manufacturing and retail functions will deliver these values with products. This will mean that processors and manufacturers will have to design products for diets that are nutritionally robust and result in lower GHG emissions, water use, and waste production. We believe the evidence base to do this exists through those sources that are freely available, and the innovations in product development can be utilised to deliver this dual goal of nutritional and environmental sustainability for 9 billion consumers. For example, the open innovation programmes developed by major food manufacturers and publishers in recent years are indicative of this emergent open-source information resource available to food product developers and companies.

Population pressures remain dominant in food security, and the transition from the rural producer to the future urban consumer in the 2050 world means there has to be an improvement in the efficiency of resource use in food processing across whole food supply chains. It is an essential component of getting food to more people using decreased inputs, and population projections provide a specific challenge to the food processing and manufacturing sectors. An essential component of realising these greater efficiencies is the integration of farming, processing, and manufacturing with regard to producing sustainable quality and quantity of food. The reduction of GHG emissions, water use and food waste will be essential, and the integration of smarter design and logistical planning in supply chains is required to do this. Marketing the sustainability agenda to consumers will help to ameliorate threats to food security that are also institutional or behavioural in nature.[3]

An understanding of the whole food system and supply chains will be required by food processors and manufacturers because consumers are changing lifestyles, expectations and demand for specific foods.[4,5] Nations that are undergoing rapid economic development, such as Brazil, Russia, India, and China, will experience large changes in manufactured food demand. This is most emphasised globally by the transition of populations living rurally to those living in urban environments. Urban living is associated with changes in how meals and diets are used, and it will change associated choice editing across all supply chains. FAOSTAT data show increased urbanisation for the global population. This scenario is specifically emphasised by the transition for China and India, where rapid urbanisation will result in dramatic changes in food demand not only in terms of volume of products but also in terms of the life-style criteria associated with those products.[6]

The future food system will result in a requirement for food processors and manufacturers to consider the following three principles:

1. The design of efficient supply chains that can deliver safe perishable foods that require efficient preservation that is dominated by the cool chain. A future food system must consider all preservation techniques that can extend the shelf life of products in urban retail environments, including ultrasound, high pressure, and irradiation treatments of food products.
2. The design of foods that provide sustainable meal planning, providing high nutrition and low environment impacts, is a focus for sustainable development. This will require an assessment of portion size and fit-for-purpose packaging so that consumers can use the appropriate amounts of food for meals and produce less domestic food waste.
3. Aligning processing and manufacturing practices with policy guidance is critical for regulatory compliance. However, the food processing and manufacturing industries hold an important supply chain position, where data sets required for footprinting products are routinely collected but not fully utilised. This provides an opportunity to lead and develop food and consumer policy in the future food system.

The integration of regulatory measures and policy-making has stimulated food manufacturing and processing companies all over the world to develop sustainability strategies that have specific focus on issues related to energy, water, waste management, and the environment.[7] These plans are driven by social, legislative, economic, and political issues that will result in food products being made with a lower energy, carbon, and water footprints. These strategies focus on the following criteria:

1. The cost of energy and water is rising and water will become scarce and vary in availability because of climate change.
2. Legislative and financial issues may restrict water use and impose restrictions on the amount of greenhouse gases a product can embody in a carbon footprint. There is also likely to be tougher legislation and/or financial costs on effluent discharges and waste generation, and all food sectors will need to be prepared to manage these changes.
3. Choice editing pressure will be imposed on the food industry by retailers and consumers for producing products that are environmentally friendly with low carbon and water footprints. Therefore, understanding how to robustly communicate processing operations to consumers is crucial to business success.
4. Companies will become more aware of their social and human rights responsibilities to their customers and the environment, and they are likely to implement 'fairer', 'greener', and 'leaner' approaches in their manufacturing operations. This is already an important issue, with many visionary companies endorsing policies on sustainability and integrating them into their mission statements.
5. Sustainable food manufacturing has been proven to have a positive impact on the profitability of these visionary companies.

Furthermore, consumer demand for sustainability criteria has increased even though most purchase decisions are clearly focused on price and quality criteria of products. An apparent opportunity for the food manufacturing and processing sector is the increased awareness of meeting energy use targets through national reporting and trading schemes that aim to decrease greenhouse gas emissions. Energy consumption reporting is critical within the food processing and manufacturing sector because energy inputs are intensive for generation of steam, hot water, and heat. The emergence of markets that support greenhouse gas emission reduction has provided sector leadership and the opportunity for companies to differentiate their products based on sustainability criteria associated with supply chains. An important aspect of developing sustainable foods remains the engagement of external and internal stakeholders for the food supply chain.

6.2 Our Conclusion

Communication of science and technology is an essential component of the practicable use of sustainability data associated with food supply chains, and this is also the target for the application of future creativity and invention in dealing with consumers. This has clear implications for managing knowledge; the communication of science in the media is often achieved by non-scientists, and the media may report accurately but not scientifically. There are naturally worlds of differences between a newspaper and a scientific journal. Furthermore, there is an extremely competitive grading and assessment system for the thousands of peer-reviewed journals within the scientific community. Does this really make any difference to what we read so long as it informs opinion, develops ideas, and keeps us up to date? Scientists in the media are as established as the term scientist itself, and current exposure to scientific communication through online and social media resources is greater than ever. A review of recent news stories from blog or social media feeds will give an idea of what drives media interest, which is constantly running 24 hours a day, every day of the year.

The food and health sector is a vibrant source of media frenzy, disruption, and current views on how large organisations develop and market new ideas and products is of almost constant interest. The emergence of new and 'old' issues, such as food security and food safety, follows on from this. Thus, media stories need not be new, but they will always catch public awareness and attention.

The quality of diet is an area where science enters the media foremost, and the relationship between diet and health is a close but overlooked one. A consideration of the cultural attitudes towards science can provide useful insights into how new ideas and technologies have developed. It is customary for science to distance itself from art and social disciplines. However, major breakthroughs in scientific study often have roots in being stimulated by cultural or market drivers. The development of management systems can help businesses provide statements of environmental and social responsibility here. They have been a significant outcome of environmental awareness. These management systems have defined aims and outcomes and can be applied to all industrial sectors. The use of LCA and environmental management system (EMS) frameworks has provided an important means for the food industry to begin to assess the impact of supply chain activity. The challenge is to make these analyses relevant to consumers by relating the science to meals and diets. These are the functional units that producers, manufacturers, and retailers must work with to ensure the delivery of sustainable products for consumers. The regulator has an important role to play because laws and enforcement can stimulate change; for example, the emergence of GHG emission trading and credits for GHG reduction has transformed energy use across the industrial arena. However, it is the incentive to improve efficiency within supply chains that provides major leaps forward because they are source of innovations and activities that will deliver what consumers want.

Developing product sustainability values alongside those of value and quality have become important to all functions of the food supply chain. This is evident in the assurance scheme and labelling revolutions that have occurred across food supply chains over the last 20 years. As an observer of this change within the agricultural and food industry, I have found that most innovations and steps towards developing sustainable products come from businesses who see opportunity in developing products that provide consumers with lifestyle choices they want, as well as the quality and price incentives. This change demands that businesses utilise both quantitative and qualitative data to guide decisions: the quantitative aspects are concerned with methodologies such as LCA, and qualitative aspects are largely concerned what consumers require, want, and how they utilise products. Using a mixed-methods approach, both quantitative and qualitative, is a necessary outcome of developing a sustainable strategy for any FMCG supply chain.

Understanding resource efficiency and the criticality of specific materials used by food supply chains will shape the form of the second Green Revolution that follows the first, which solved many of the production issues that were the cause of food insecurity in the twentieth century. We have seen the emergence of assurance and traceability schemes in the twentieth-first century and their establishment has strengthened the values associated with foods. There remain significant challenges to providing food security, and it is clear that what, and how, the consumer uses ingredients and foods is crucial to delivering sustainability and security to the global food system. The development of the current food system has established trade organisations and agreements that will respond to security and criticality issues more efficiently than ever before and provide a deeper understanding of criticality within supply chains. Developing responsibility across the food system will mean there is an increasing requirement to understand what consumers require. This will result in far more creative ways to handle consumer data, and this will require integration of qualitative and quantitative analyses. Increasingly developments that solve sustainability challenges are being made outside of government actions and incentives because there is a very clear business case to achieve security and sustainability. The unit of this change is the meal we eat three or four times a day, the recipes we prepare and the foods we order. Small changes in obtaining the appropriate protein intake, reducing food waste and getting the correct nutritional balance associated with this plate of food will provide solutions to the challenges that currently face us. It is hoped that this book goes some way to highlight how these changes are going to be made, managed, and measured by humankind's future food system.

References

1 Chouinard, Y. (2006). *Let my people go surfing: the education of a reluctant businessman*. Penguin.

2 Chouinard, Y., Ellison, J., & Ridgeway, R. (2011). The sustainable economy. *Harvard Business Review*, 89(10), 52–62.

3 Bánáti, D. (2011). Consumer response to food scandals and scares. *Trends in Food Science & Technology*, 22(2), 56–60.

4 Popkin, B. M. (2001). The nutrition transition and obesity in the developing world. *The Journal of Nutrition*, 131(3), 871S–873S.

5 Drewnowski, A., & Popkin, B. M. (1997). The nutrition transition: new trends in the global diet. *Nutrition Reviews*, 55(2), 31–43.

6 Popkin, B. M., & Gordon-Larsen, P. (2004). The nutrition transition: worldwide obesity dynamics and their determinants. *International Journal of Obesity*, 28, S2–S9.

7 Dauvergne, P., & Lister, J. (2012). Big brand sustainability: governance prospects and environmental limits. *Global Environmental Change*, 22(1), 36–45.

Index

Global Food Security and Supply, First Edition. Wayne Martindale.
© 2015 John Wiley & Sons, Ltd. Published 2015 by John Wiley & Sons, Ltd.

Food Science and Technology Books

WILEY Blackwell

GENERAL FOOD SCIENCE & TECHNOLOGY, ENGINEERING AND PROCESSING

Title	Author	ISBN
Food Texture Design and Optimization	Dar	9780470672426
Nano- and Microencapsulation for Foods	Kwak	9781118292334
Extrusion Processing Technology: Food and Non-Food Biomaterials	Bouvier	9781444338119
Food Processing: Principles and Applications, 2nd Edition	Clark	9780470671146
The Extra-Virgin Olive Oil Handbook	Peri	9781118460450
Mathematical and Statistical Methods in Food Science and Technology	Granato	9781118433683
The Chemistry of Food	Velisek	9781118383841
Dates: Postharvest Science, Processing Technology and Health Benefits	Siddiq	9781118292372
Resistant Starch: Sources, Applications and Health Benefits	Shi	9780813809519
Statistical Methods for Food Science: Introductory 2nd Edition	Bower	9781118541647
Formulation Engineering of Foods	Norton	9780470672907
Practical Ethics for Food Professionals: Research, Education and the Workplace	Clark	9780470673430
Edible Oil Processing, 2nd Edition	Hamm	9781444336849
Bio-Nanotechnology: A Revolution in Food, Biomedical and Health Sciences	Bagchi	9780470670378
Dry Beans and Pulses : Production, Processing and Nutrition	Siddiq	9780813823874
Genetically Modified and non-Genetically Modified Food Supply Chains: Co-Existence and Traceability	Bertheau	9781444337785
Food Materials Science and Engineering	Bhandari	9781405199223
Handbook of Fruits and Fruit Processing, second edition	Sinha	9780813808949
Tropical and Subtropical Fruits: Postharvest Physiology, Processing and Packaging	Siddiq	9780813811420
Food Biochemistry and Food Processing, 2nd Edition	Simpson	9780813808741
Dense Phase Carbon Dioxide	Balaban	9780813806495
Nanotechnology Research Methods for Food and Bioproducts	Padua	9780813817316
Handbook of Food Process Design, 2 Volume Set	Ahmed	9781444330113
Ozone in Food Processing	O'Donnell	9781444334425
Food Oral Processing	Chen	9781444330120
Food Carbohydrate Chemistry	Wrolstad	9780813826653
Organic Production & Food Quality	Blair	9780813812175
Handbook of Vegetables and Vegetable Processing	Sinha	9780813815411

FUNCTIONAL FOODS, NUTRACEUTICALS & HEALTH

Title	Author	ISBN
Antioxidants and Functional Components in Aquatic Foods	Kristinsson	9780813813677
Food Oligosaccharides: Production, Analysis and Bioactivity	Moreno-Fuentes	9781118426494
Novel Plant Bioresources: Applications in Food, Medicine and Cosmetics	Gurib-Fakim	9781118460610
Functional Foods and Dietary Supplements: Processing Effects and Health Benefits	Noomhorm	9781118227879
Food Allergen Testing: Molecular, Immunochemical and Chromatographic Techniques	Siragakis	9781118519202
Bioactive Compounds from Marine Foods: Plant and Animal Sources	Hernández-Ledesma	9781118412848
Bioactives in Fruit: Health Benefits and Functional Foods	Skinner	9780470674970
Marine Proteins and Peptides: Biological Activities and Applications	Kim	9781118375068
Dried Fruits: Phytochemicals and Health Effects	Alasalvar	9780813811734
Handbook of Plant Food Phytochemicals	Tiwari	9781444338102
Analysis of Antioxidant-Rich Phytochemicals	Xu	9780813823911
Phytonutrients	Salter	9781405131513
Coffee: Emerging Health Effects and Disease Prevention	Chu	9780470958780
Functional Foods, Nutraceuticals & Disease Prevention	Paliyath	9780813824536
Nondigestible Carbohydrates and Digestive Health	Paeschke	9780813817620
Bioactive Proteins and Peptides as Functional Foods and Nutraceuticals	Mine	9780813813110
Probiotics and Health Claims	Kneifel	9781405194914
Functional Food Product Development	Smith	9781405178761

INGREDIENTS

Title	Author	ISBN
Fats in Food Technology, 2nd Edition	Rajah	9781405195423
Processing and Nutrition of Fats and Oils	Hernandez	9780813827674
Stevioside: Technology, Applications and Health	De	9781118350669
The Chemistry of Food Additives and Preservatives	Msagati	9781118274149
Sweeteners and Sugar Alternatives in Food Technology, 2nd Edition	O'Donnell	9780470659687
Hydrocolloids in Food Processing	Laaman	9780813820767
Natural Food Flavors and Colorants	Attokaran	9780813821108
Handbook of Vanilla Science and Technology	Havkin-Frenkel	9781405193252
Enzymes in Food Technology, 2nd edition	Whitehurst	9781405183666
Food Stabilisers, Thickeners and Gelling Agents	Imeson	9781405132671
Glucose Syrups - Technology and Applications	Hull	9781405175562
Dictionary of Flavors, 2nd edition	DeRovira	9780813821351

FOOD SAFETY, QUALITY AND MICROBIOLOGY

Title	Author	ISBN
Practical Food Safety	Bhat	9781118474600
Food Chemical Hazard Detection	Wang	9781118488591
Food Safety for the 21st Century	Wallace	9781118897980
Guide to Foodborne Pathogens, 2nd Edition	Labbe	9780470671429
Improving Import Food Safety	Ellefson	9780813808772
Food Irradiation Research and Technology, 2nd Edition	Fan	9780813802091
Food Safety: The Science of Keeping Food Safe	Shaw	9781444337228

For further details and ordering information, please visit www.wiley.com/go/food

Food Science and Technology from Wiley Blackwell

Decontamination of Fresh and Minimally Processed Produce	Gomez-Lopez	9780813823843
Progress in Food Preservation	Bhat	9780470655856
Food Safety for the 21st Century: Managing HACCP and Food Safety throughout the Global Supply Chain	Wallace	9781405189118
The Microbiology of Safe Food, 2nd edition	Forsythe	9781405140058

SENSORY SCIENCE, CONSUMER RESEARCH & NEW PRODUCT DEVELOPMENT

Olive Oil Sensory Science	Monteleone	9781118332528
Quantitative Sensory Analysis: Psychophysics, Models and Intelligent Design	Lawless	9780470673461
Product Innovation Toolbox: A Field Guide to Consumer Understanding and Research	Beckley	9780813823973
Sensory and Consumer Research in Food Product Design and Dev, 2nd Ed	Moskowitz	9780813813660
Sensory Evaluation: A Practical Handbook	Kemp	9781405162104
Statistical Methods for Food Science	Bower	9781405167642
Concept Research in Food Product Design and Development	Moskowitz	9780813824246
Sensory and Consumer Research in Food Product Design and Development	Moskowitz	9780813816326

FOOD INDUSTRY SUSTAINABILITY & WASTE MANAGEMENT

Food and Agricultural Wastewater Utilization and Treatment, 2nd Edition	Liu	9781118353974
Sustainable Food Processing	Tiwari	9780470672235
Food and Industrial Bioproducts and Bioprocessing	Dunford	9780813821061
Handbook of Sustainability for the Food Sciences	Morawicki	9780813817354
Sustainability in the Food Industry	Baldwin	9780813808468
Lean Manufacturing in the Food Industry	Dudbridge	9780813810072

FOOD LAWS & REGULATIONS

Guide to US Food Laws and Regulations, 2nd Edition	Curtis	9781118227787
Food and Drink - Good Manufacturing Practice: A Guide to its Responsible Management (GMP6), 6th Edition	Manning	9781118318201
The BRC Global Standard for Food Safety: A Guide to a Successful Audit, 2nd Edition	Kill	9780470670651
Food Labeling Compliance Review, 4th edition	Summers	9780813821818

DAIRY FOODS

Lactic Acid Bacteria: Biodiversity and Taxonomy	Holzapfel	9781444333831
From Milk By-Products to Milk Ingredients: Upgrading the Cycle	de Boer	9780470672228
Milk and Dairy Products as Functional Foods	Kanekanian	9781444336832
Milk and Dairy Products in Human Nutrition: Production, Composition and Health	Park	9780470674185
Manufacturing Yogurt and Fermented Milks, 2nd Edition	Chandan	9781119967088
Sustainable Dairy Production	de Jong	9780470655849
Advances in Dairy Ingredients	Smithers	9780813823959
Membrane Processing: Dairy and Beverage Applications	Tamime	9781444333794
Analytical Methods for Food and Dairy Powders	Schuck	9780470655986
Dairy Ingredients for Food Processing	Chandan	9780813817460
Processed Cheeses and Analogues	Tamime	9781405186421
Technology of Cheesemaking, 2nd edition	Law	9781405182980

SEAFOOD, MEAT AND POULTRY

Seafood Processing: Technology, Quality and Safety	Boziaris	9781118346211
Should We Eat Meat? Evolution and Consequences of Modern Carnivory	Smil	9781118278727
Handbook of Meat, Poultry and Seafood Quality, second edition	Nollet	9780470958322
The Seafood Industry: Species, Products, Processing, and Safety , 2nd Edition	Granata	9780813802589
Organic Meat Production and Processing	Ricke	9780813821269
Handbook of Seafood Quality, Safety and Health Effects	Alasalvar	9781405180702

BAKERY & CEREALS

Oats Nutrition and Technology	Chu	9781118354117
Cereals and Pulses: Nutraceutical Properties and Health Benefits	Yu	9780813818399
Whole Grains and Health	Marquart	9780813807775
Gluten-Free Food Science and Technology	Gallagher	9781405159159
Baked Products - Science,Technology and Practice	Cauvain	9781405127028

BEVERAGES & FERMENTED FOODS/BEVERAGES

Encyclopedia of Brewing	Boulton	9781405167444
Sweet, Reinforced and Fortified Wines: Grape Biochemistry, Technology and Vinification	Mencarelli	9780470672242
Technology of Bottled Water, 3rd edition	Dege	9781405199322
Wine Flavour Chemistry, 2nd edition	Bakker	9781444330427
Wine Quality: Tasting and Selection	Grainger	9781405113663

PACKAGING

Handbook of Paper and Paperboard Packaging Technology, 2nd Edition	Kirwan	9780470670668
Food and Beverage Packaging Technology, 2nd edition	Coles	9780813812748
Food and Package Engineering	Morris	9780813814797
Modified Atmosphere Packaging for Fresh-Cut Fruits and Vegetables	Brody	9780813812748

For further details and ordering information, please visit www.wiley.com/go/foo